Mechanisms and Machine Science

Volume 30

Series editor
Marco Ceccarelli, Cassino, Italy

More information about this series at http://www.springer.com/series/8779

Erwin-Christian Lovasz
Gondi Kondaiah Ananthasuresh
Burkhard Corves • Victor Petuya
Editors

Microactuators and Micromechanisms

Proceedings of MAMM 2014, Timisoara, Romania, October 2-4, 2014

Springer

Editors

Erwin-Christian Lovasz
Department of Mechatronics
University Politehnica Timisoara
Timisoara, Romania

Gondi Kondaiah Ananthasuresh
Mechanical Engineering
Indian Institute of Science
Bangalore, India

Burkhard Corves
Institut für Getriebetechnik und Maschinendynamik
RWTH Aachen University
Aachen, Germany

Victor Petuya
Department of Mechanical Engineering
University of the Basque Country
Bilbao, Spain

ISSN 2211-0984 ISSN 2211-0992 (electronic)
Mechanisms and Machine Science
ISBN 978-3-319-15861-7 ISBN 978-3-319-15862-4 (eBook)
DOI 10.1007/978-3-319-15862-4

Library of Congress Control Number: 2015937012

Springer Cham Heidelberg New York Dordrecht London

Printed on acid-free paper

Springer International Publishing AG Switzerland is part of Springer Science+Business Media (www.springer.com)

Committees

Workshop Chairman
Erwin-Christian Lovasz (Romania)

Workshop Co-Chairmen
Gondi Kondaiah Ananthasuresh (India)
Burkhard Corves (Germany)
Victor Petuya (Spain)

International Scientific Committee
Brian Jensen (USA)
Gondi Kondaiah Ananthasuresh (India)
Burkhard Corves (Germany)
Amitabha Ghosh (India)
Antoni Gronowicz (Poland)
Karl-Heinz Modler (Germany)
Victor Petuya (Spain)
Anupam Saxena (India)
Ionuţ Doroftei (Romania)
Hidetsugu Terada (Japan)
Srikar Vengalattore (Canada)
Niels Modler (Germany)
Yao Yan-An (P.R. China)
Lena Zentner (Germany)

Local Organising Committee
Corina Mihaela Gruescu
Inocenţiu Maniu
Valentin Ciupe
Iosif Cărăbaş
Nicolae Mircea Dehelean
Valer Dolga

Florina Pop
Cristian Pop
Dan Teodor Mărgineanu
Cristian Emil Moldovan
Eugen Sever Zăbavă

With the Support of the Romanian IFToMM National Committee – ARoTMM
Ionuţ Doroftei
Doina Pâslă
Erwin-Christian Lovasz
Eugen Merticaru

Preface

The third Conference on Microactuators and Micromechanisms, MAMM-2014, was organized by the Department of Mechatronics at the Faculty of Mechanical Engineering, Politehnica University of Timisoara, under the patronage of the International Federation for the Promotion of Mechanism and Machine Science (IFToMM) Technical Committees for Linkages and Mechanical Controls and for Micromachines.

The Conference aims to bring together researchers and students who develop their work in disciplines associated with micromechanisms and microactuators in a friendly, colleague-like, and collaborative environment. MAMM-2014 offers a great opportunity for scientists all over the world to present their achievements, exchange innovative ideas, and create solid international links. The scientific event was meant to gather the novelty and originality in machine science and to depict the trend of this important and creative field.

The topics proposed for the Conference are Microactuators and micro-assembly, Micro sensors involving movable solids, Micro-opto-mechanical device, Mechanical tools for cell and tissue studies, Micromanipulation and micro-stages, Microscale flight and swimming, Micro-robotics and surgical tools, Micron-scale power generation, Miniature manufacturing machines, Micromechatronics and micromechanisms, Biomechanics micro and nano scales and Control issues in microsystems.

We would like to express our grateful thanks to IFToMM, to the Romania IFToMM National Committee (ARoTMM), and to the members of the International Scientific Committee of MAMM-2014.

We appreciate the effort of the reviewers gathered in the International Scientific Committee of MAMM-2014. They spent time on a serious work of evaluation and improvement guidance, meant to assure high quality in all papers.

We thank the authors, who contributed with valuable papers on different subjects covering the wide scientific fiels of Mechanisms and Machine Science.

We thank the Department of Mechatronics at the Faculty of Mechanical Engineering, Politehnica University of Timisoara, for hosting the scientific event and supporting all associated activities. The third Conference on Microactuators and Micromechanisms became a reality due to a hardworking local organizing team: Gruescu Corina Mihaela, Inocenţiu Maniu, Valentin Ciupe, Iosif Cărăbaş, Nicolae Mircea Dehelean, Valer Dolga, Florina Pop, Cristian Pop, Dan Teodor Mărgineanu, Cristian Emil Moldovan, and Eugen Sever Zăbavă.

Last but not least, we are grateful to the staff at Springer Publishing for their excellent technical and editorial support.

Timisoara, Romania
September 2014

Erwin-Christian Lovasz
Gondi Kondaiah Ananthasuresh
Burkhard Corves
Victor Petuya

Contents

A Compliant Mechanism as Rocker Arm with Spring Capability for Precision Engineering Applications 1
L. Hartmann and L. Zentner

Steering and Non-steering Crawling Tetrahedral Micro-mechanisms 9
D. Mărgineanu, E.-C. Lovasz, K.-H. Modler, and C.M. Gruescu

New Kinematic Designs of Flexure Hinge Based 3 DoF Translational Micromanipulators 23
I. Prause, D. Schoenen, and B. Corves

Development of Adaptive Compliant Gripper Finger with Embedded Actuators 33
A. Milojević and N.D. Pavlović

Overview and Classification of Flexure Hinge Based Micromanipulators 51
D. Schoenen, I. Prause, S. Palacios, and B. Corves

Axis Cross-Coupling Reduction on a High Bandwidth XY Flexure Stage 61
A. Ruiz, F.J. Campa, O. Altuzarra, V. Petuya, C. Pinto, and A. Hernández

Some Structural and Kinematic Characteristics of Micro Walking Robots 73
Adr. Comanescu, I. Dugaesescu, and D. Comanescu

Development of a 3-DOF Compliant Robotic Local Structure with Large Twist Angle 87
S. Kurtenbach, J. Siebrecht, D. Schoenen, M. Hüsing, and B. Corves

A Hexapod Walking Micro-robot with Artificial Muscles 99
I. Doroftei

Side Chain Kinematics Simulation on Protein Conformational Changes 121
Mikel Diez, Victor Petuya, Mónica Urizar, Oscar Altuzarra, and Alfonso Hernández

Design and Fabrication of Millimeter-Scale Crossed-Cylinder Wrist Mechanism with Two Degrees of Freedom 133
Brian D. Jensen, Jordan Tanner, Bryce Edmondson, Clayton Grames, Spencer P. Magleby, and Larry L. Howell

Conception of a Mechanical System for Rehabilitation of Hand Function for Use in Medical Training Therapy 143
M. Feierabend and L. Zentner

Modeling, Parametric Synthesis and Characterization of Bellow-Type Pneumatic Micro-actuator 159
J. Prateek and G. Prasanna

Stability Analysis of Semi-kinematic Mountings used in Modular Reconfigurable Micro Factory Testbed 169
Mounika Katragadda, Aneissha Chebolu, and Nagahanumaiah

Multibody System Simulation of Hysteresis Effect by Micro Textile-Reinforced Compliant Mechanisms with Piezo-electric Actuators 177
D. Mărgineanu, E.-C. Lovasz, and N. Modler

Electrical Performance of Electrically Conductive Silicone Rubber in Dependence on Tensile Load According to Various Parameters 191
M. Issa and L. Zentner

Contributors

O. Altuzarra Department of Mechanical Engineering, University of the Basque Country UPV/EHU, Bilbao, Spain

F.J. Campa Department of Mechanical Engineering, University of the Basque Country UPV/EHU, Bilbao, Spain

Aneissha Chebolu Micro Systems Technology Laboratories, CSIR-Central Mechanical Engineering Research Institute, Durgapur, India

Adr. Comanescu Department of Mechanisms and Robots Theory, University Politehnica of Bucharest, Bucharest, Romania

D. Comanescu Department of Mechanisms and Robots Theory, University Politehnica of Bucharest, Bucharest, Romania

Burkhard Corves Institut für Getriebetechnik und Maschinendynamik, RWTH Aachen University, Aachen, Germany

Mikel Diez Department of Mechanical Engineering, University of the Basque Country UPV/EHU, Bilbao, Spain

I. Doroftei Mechanical Engineering Faculty, Theory of Mechanisms and Robotics Department, "Gheorghe Asachi" Technical University of Iasi, Iasi, Romania

I. Dugaesescu Department of Mechanisms and Robots Theory, University Politehnica of Bucharest, Bucharest, Romania

Bryce Edmondson Department of Mechanical Engineering, Brigham Young University, Provo, UT, USA

M. Feierabend Department of Mechanical Engineering, Mechanism Technology Group, Technische Universität Ilmenau, Ilmenau, Germany

Clayton Grames Department of Mechanical Engineering, Brigham Young University, Provo, UT, USA

C.M. Gruescu Department of Mechatronics, University Politehnica Timisoara, Timisoara, Romania

L. Hartmann Department of Mechanical Engineering, Mechanism Technology Group, Technische Universität Ilmenau, Ilmenau, Germany

Alfonso Hernández Department of Mechanical Engineering, University of the Basque Country UPV/EHU, Bilbao, Spain

Larry L. Howell Department of Mechanical Engineering, Brigham Young University, Provo, UT, USA

M. Hüsing Institut für Getriebetechnik und Maschinendynamik, RWTH Aachen University, Aachen, Germany

M. Issa Department of Mechanical Engineering, Mechanism Technology Group, Technische Universität Ilmenau, Ilmenau, Germany

Brian D. Jensen Department of Mechanical Engineering, Brigham Young University, Provo, UT, USA

Mounika Katragadda Micro Systems Technology Laboratories, CSIR-Central Mechanical Engineering Research Institute, Durgapur, India

S. Kurtenbach Institut für Getriebetechnik und Maschinendynamik, RWTH Aachen University, Aachen, Germany

E.-C. Lovasz Department of Mechatronics, University Politehnica Timisoara, Timisoara, Romania

Spencer P. Magleby Department of Mechanical Engineering, Brigham Young University, Provo, UT, USA

D. Mărgineanu Department of Mechatronics, University Politehnica Timisoara, Timisoara, Romania

A. Milojević Faculty of Mechanical Engineering, Department for Mechatronics and Control, University of Niš, Niš, Serbia

N. Modler Institut für Leichtbau und Kunststofftechnik, Technical University Dresden, Dresden, Germany

K.-H. Modler Institut für Leichtbau und Kunststofftechnik, Technical University Dresden, Dresden, Germany

Nagahanumaiah Micro Systems Technology Laboratories, CSIR-Central Mechanical Engineering Research Institute, Durgapur, India

S. Palacios Institut für Getriebetechnik und Maschinendynamik, RWTH Aachen University, Aachen, Germany

N.D. Pavlović Faculty of Mechanical Engineering, Department for Mechatronics and Control, University of Niš, Niš, Serbia

Victor Petuya Department of Mechanical Engineering, University of the Basque Country UPV/EHU, Bilbao, Spain

C. Pinto Department of Mechanical Engineering, University of the Basque Country UPV/EHU, Bilbao, Spain

G. Prasanna Department of Mechanical Engineering, Indian Institute of Technology Bombay, Mumbai, India

J. Prateek Department of Mechanical Engineering, Indian Institute of Technology Bombay, Mumbai, India

I. Prause Institut für Getriebetechnik und Maschinendynamik, RWTH Aachen University, Aachen, Germany

A. Ruiz Department of Mechanical Engineering, University of the Basque Country UPV/EHU, Bilbao, Spain

D. Schoenen Institut für Getriebetechnik und Maschinendynamik, RWTH Aachen University, Aachen, Germany

J. Siebrecht Institut für Getriebetechnik und Maschinendynamik, RWTH Aachen University, Aachen, Germany

Jordan Tanner Department of Mechanical Engineering, Brigham Young University, Provo, UT, USA

Mónica Urizar Department of Mechanical Engineering, University of the Basque Country UPV/EHU, Bilbao, Spain

L. Zentner Department of Mechanical Engineering, Mechanism Technology Group, Technische Universität Ilmenau, Ilmenau, Germany

A Compliant Mechanism as Rocker Arm with Spring Capability for Precision Engineering Applications

L. Hartmann and L. Zentner

Abstract In precision engineering escapement and ratchet mechanism are often used to keep an element in a non-operating state e.g. in chronometers. Such mechanisms have a pivoted and spring-loaded pawl locking a ratchet. For miniaturized applications reducing the number of elements and functional integration is recommended. Spring clip mechanisms as rocker arms combine the rocker itself with a spring. Therefor a special formed wire is mounted in a frame by two shifted revolute joints. The spring capability is once determined by the geometry and the material and twice by the shifting of the axes and the deflection of the clip. In this contribution the force-deflection behavior of such mechanisms for different shiftings is focused utilizing the finite element method. Spring clip mechanisms realize a wide range of motion and due to a shifting of axes and the asymmetry the force-deflection behavior is influenced. Finally some precision engineering devices with a spring clip mechanism are suggested.

Keywords Compliant mechanism • Precision engineering • Rocker arm • Spring • Spring clip mechanism

1 Introduction

Designing scientific and measuring instruments or highly accurate devices, such as chronometers, medical and optical apparatus, is a part of precision engineering. Usually the motion of parts and assemblies in precision engineering is characterized by high velocities but the inertia has to be exceptionally low (Bischoff 1959; Hildebrand 1967; Schilling 1989). To reach a high precision and a minor inertia, the components and machine elements of such devices have small dimensions and consequently a low mass. So, downsizing of these elements is recommended and this is a reason for integration of functions. Additionally, manufacturing and assembling of miniaturized and function integrated components is more difficult than macro machine elements.

L. Hartmann (✉) • L. Zentner
Department of Mechanical Engineering, Mechanism Technology Group,
Technische Universität Ilmenau, Ilmenau, Germany
e-mail: lars.hartmann@tu-lmenau.de; lena.zentner@tu-ilmenau.de

E.-C. Lovasz et al. (eds.), *Microactuators and Micromechanisms*, Mechanisms and Machine Science 30, DOI 10.1007/978-3-319-15862-4_1

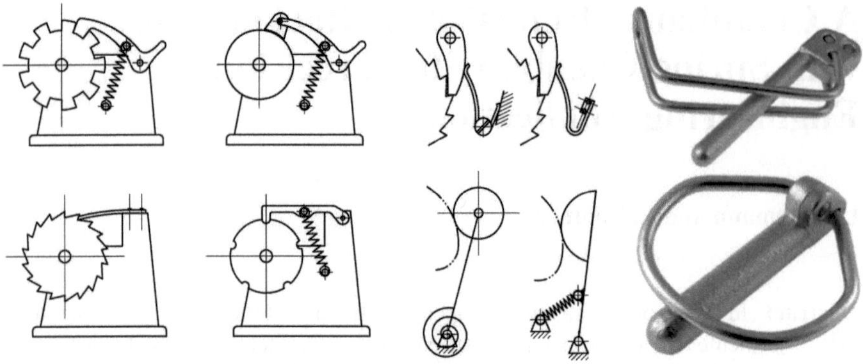

Fig. 1 Escapement and ratchet mechanisms with spring-loaded pawls, cam mechanism with different spring loaded rocker arms, linchpins as examples for spring clip mechanisms (Bischoff 1959; Hildebrand 1967; RÜBIG GmbH & Co. 2013; Schilling 1989)

In several precision engineering devices escapement and ratchet mechanisms are used to keep an element in a non-operating state, Fig. 1. Therefor a ratchet is locked by a pawl, which is usually pivoted and spring-loaded. Also in cam mechanisms springs are used to keep the mechanism elements in contact, Fig. 1. The spring in each application can be a coil spring or a bending spring. For a bending spring like a cantilever beam the range of motion is strictly limited because of the mechanical stress. To realize a wide range of motion a rocker arm is loaded by tension or spiral springs but these are additional parts in construction. With a look at mechanisms with pivotally mounted rockers, especially escapement and ratchet mechanism, we introduce a spring clip mechanism as a rocker arm with spring capability for precision engineering applications. Spring clip mechanisms as compliant mechanisms consist of a specially shaped spring clip (wire bracket) with bending and torsional capability which is mounted in a rigid frame by two revolute joints with shifted axes, Fig. 1. The spring capability is determined by the geometry, the material, the distance between the two shifted axes (displacement) and the deflection of the clip. We aim to investigate the force-deflection behavior of such a mechanism for different displacements.

2 Material and Methods

Analyzing spring clip mechanisms and to investigate their stiffness behavior mechanisms we suggest a model to use the finite element method.

2.1 Modeling

An essential part in the investigation of a spring clip mechanism is the modeling. Generally, there are several specially formed spring clips, each adapted for one special application. Because it is impossible to focus on all of them, one particular spring clip is chosen for the investigation. The spring clip mechanism we use for studying has features and restrictions as follows (Hartmann and Zentner 2013):

- it is built up of three bending and torsion flexible, straight legs,
- in undeformed state it is planar, meaning that all bar axes are within one plane and all legs are perpendicular to each other,
- asymmetry occurs because of the different lengths of the two parallel legs,
- slender legs with small and constant circular cross-sections throughout the length of all legs ,
- plane sections remain plane and normal to the deflected neutral axis,
- the material is isotropic and homogeneous and obeys Hooke's law,
- friction in all hinges is neglected and gravity is not observed.

2.2 *Finite Element Analysis*

To investigate the static structural deformation of a spring clip, the finite element method (FEM) is well-suited. Using the software package ANSYS® Workbench 14.5, it is easy to handle a large number of analyses with different analysis settings, element types, boundary conditions, displacements of one joint and deflections of the spring clip. Despite all simplifications modeling the spring clip mechanism by elastic 3D-beam elements (*beam4*) to apply the Euler-Bernoulli theory will produce wrong results, because of the restrictions of infinitesimal strain and small rotations. Instead we have to use a 3D-solid tetrahedron element with a linear element shape function, the solid 185-element and large deflections have to be enabled in analysis settings. Only this non-linear approach generates useful results for the force-deflection behavior of the spring clip mechanism. The hinges A and B are modeled as joints with a rotational motion around the *z*-axis and the displacement is applied to the hinge B along the *y*-axis. Subsequent to this first load step the free end of the spring clip will deflect and the force reaction at one point of the free end is analyzed. Figure 2 visualizes the described FEM model and the parameters.

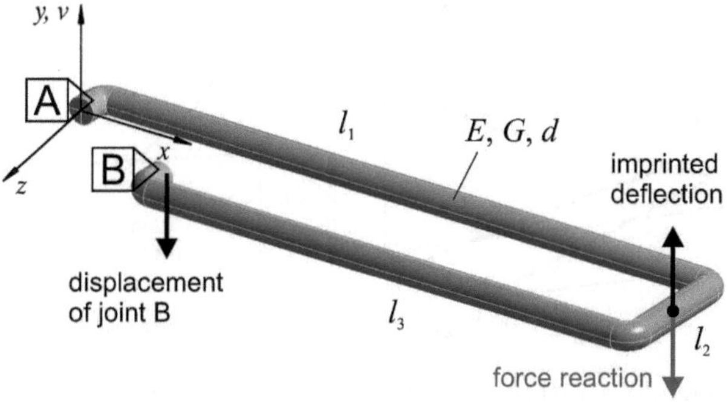

length leg 1	$l_1 = 5$ mm
length leg 2	$l_2 = 1$ mm
length leg 3	l_3 variable
diameter	d variable
Young's modulus	$E = 206$ kN/mm^2
shear modulus	$G = 78$ kN/mm^2

Fig. 2 FEM model for analyses of an asymmetrical spring clip mechanism and parameters (Hartmann and Zentner 2013)

The length of leg 3 is $l_3 = 4$ mm and $l_3 = 3$ mm to investigate the effect of asymmetry of the spring clip to the force-deflection behavior. Diameters of the wire are $d = 0.2$ mm and $d = 0.4$ mm to ascertain its effect on the force-deflection behavior. Fixing 0.2 and 0.4 mm for the displacements of hinge B and the imprinting the maximum deflection (same value as the resulting deflection due to displacement of hinge B but in the opposite direction) in the second load step the force-deflection behavior of the mechanism for different displacements is evaluable.

3 Results

Firstly, the deformed shapes of asymmetrical spring clips without a displacement of hinge B are presented in Fig. 3. The maximum imprinted deflection at the tip is ±1 mm and all legs are nearly straight. According to the deformed shapes the force reactions at the tip are evaluated. The force-deflection behavior is non-linear, in consequence the spring rate is not constant. If there is no imprinted deflection on the tip no force is generated. Deflection and force reaction go into opposite directions, meaning that a positive deflection causes a negative force reaction. Finally, increasing the diameter of the wire offers higher forces.

Applying a displacement at hinge B in negative y-direction causes a resulting deflection of the spring clip also in negative y-direction. In the case of a nearly symmetrical spring clip ($l_3 = 4$ mm) this deflection of the center tip point is

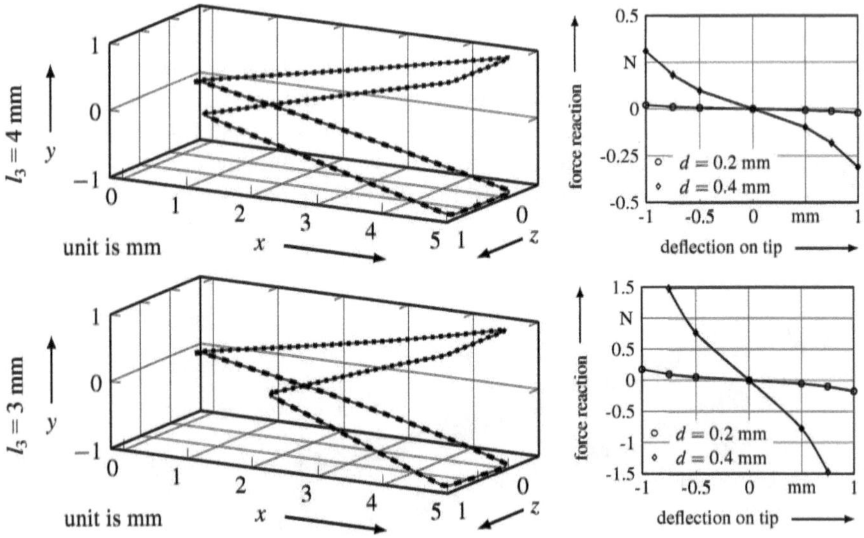

Fig. 3 Deformed shapes and force-deflection behavior without any displacement of hinge B; *left* – deformed shapes of spring clip; *right* – force-deflection behavior

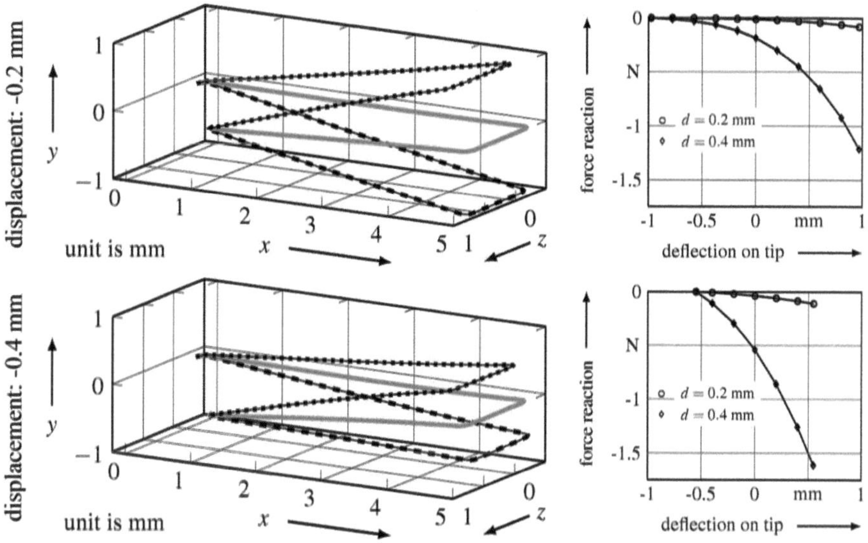

Fig. 4 Deformed shapes and force-deflection behavior with displacement of hinge B for $l_3 = 4$ mm; *left* – deformed shapes of spring clip; *right* – force-deflection behavior

−0.97 mm for a displacement of −0.2 and −0.55 mm for −0.4 mm. The dashed lines in Fig. 4 show this state, where no force reaction is generated. Deformed shape of the clip for an imprinted deflection bringing the tip center point to $y = 0$ is represented by the gray lines in Fig. 4. The dotted lines in Fig. 4 characterize the deformation, if the center tip point is moved to $y = +0.97$ mm or +0.55 mm, depending on the displacement of the hinge B. Contrary to the expectation the minor displacement of hinge B effects a larger deflection than the major one. Again the force-deflection behavior is non-linear but for a deflection $v = 0$ there is a force reaction. The force increases for higher imprinted deflections and major diameters.

According to Figs. 4 and 5 illustrates the effects for a more asymmetrical spring clip ($l_3 = 3$ mm). As expected, major displacements of hinge B realize higher resulting deflections than miner ones (see dotted and dashed lines in Fig. 5). For major displacements and large imprinted deflections the legs are not straight anymore. The force reactions in this case are high in comparison to the nearly symmetrical cases.

In all cases the trajectory of the points of leg 2 is nearly a circular arc in *x*-*y*-plane. The mechanical stress is low due to the rotation of the leg 1 and leg 3 around the *z*-axes. But for major displacements of hinge B and short leg lengths l_3 it will raise due to increasing bending and torsion torques. If the imprinted deflection is too large, bending around *y*-axis has more and more influence on the deformation and the shape deflects to positive *z*-direction (site of shorter leg 3).

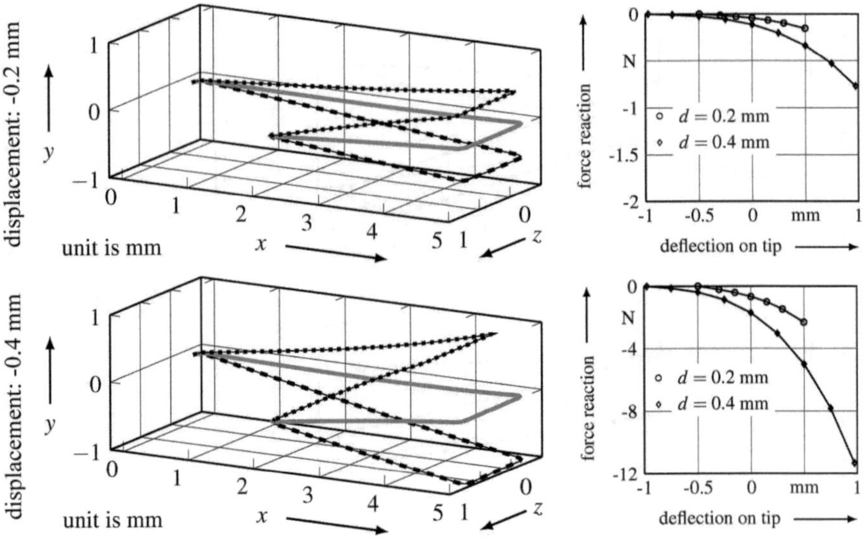

Fig. 5 Deformed shapes and force-deflection behavior with displacement of hinge B for $l_3 = 3$ mm; *left* – deformed shapes of spring clip; *right* – force-deflection behavior

4 Discussion

The results presented in chapter "New kinematic designs of flexure hinge based 3 DoF translational micromanipulators" show that the shifting of the axes (displacement of hinge B) and the asymmetry of a spring clip mechanism influences the force-deflection behavior. In all cases we looked at, the force-deflection behavior is non-linear. Changing the geometry and material parameters of the spring clip mechanisms other force-deflection behaviors can be achieved. Especially for a non-constant wire diameter over the arc length of all legs the spring rate is adjustable. Finally, the wide range of motion, the spring capability and the adjustable spring force are advantages for precision engineering applications. To realize a rocker arm with an angular motion the leg 3 as output element is preferred because of the low bending deformation.

There are several forms of spring clips for specific applications in precision engineering. Using spring clip mechanisms as rocker arms with spring capability reduces the number of parts because of the integration of lever and spring. Of course scaling the geometry and miniaturization of such mechanisms to adapt them to special applications is possible and their manufacturing is easy. Figure 6 visualizes some applications for these mechanisms.

Generally, solving the problem is not trivial even if the modeling seams to offer the usage of a linear theory. Due to the large deflections this highly non-linear problem can be solved by the finite element method but also with some restrictions. To reach a convergent solution and acceptable results the FEM model has to be well adapted at real environmental conditions.

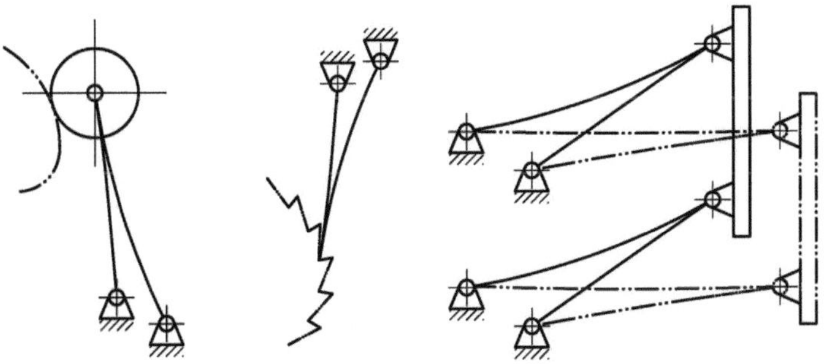

Fig. 6 Application of spring clip mechanisms in precision engineering; *left* – cam mechanism with spring clip mechanism as rocker arm; *center* – spring clip mechanism as pawl in a ratchet mechanism; *right* – parallel spring with a wide range of motion due to spring clip mechanisms

5 Conclusions

A spring clip mechanism with shifted hinges was analyzed using the finite element method. The shifting of axes (displacement) has an influence to the force-deflection behavior. Because of the integration of functions – rocker arm with spring capability – these compliant mechanisms establish new precision engineering applications.

Acknowledgments In memoriam, the authors would like to thank gratefully Prof. Gerhard Bögelsack.

References

Bischoff W (1959) Das Grundprinzip als Schlüssel zur Systematisierung, vol 5. Wiss. Zeitschr. d. TH Ilmenau, p 199

Hartmann L, Zentner L (2013) Über die Wirkung einer mechanischen Vorspannung auf die Deformation eines asymmetrische Federbügel-mechanismus. In: Zentner L (Hrsg) 10. Kolloquium Getriebetechnik 2013. Universitätsverlag Ilmenau, Ilmenau, pp 373–390

Hildebrand S (1967) Feinmechanische Bauelemente. VEB Verlag Technik Berlin, Berlin, pp 667–690

RÜBIG GmbH & Co. (2013) KG: RÜBIG Klappstecker. URL: http://www.rubig.com/index.cfm?seite = klappstecker-st&sprache = DE. Accessed 18th June 2013

Schilling M (1989) Gehemme und Gesperre. In: Krause W (ed) Konstruktionselemente der Feinmechanik. VEB Verlag Technik Berlin, Berlin, pp 446–462

Steering and Non-steering Crawling Tetrahedral Micro-mechanisms

D. Mărgineanu, E.-C. Lovasz, K.-H. Modler, and C.M. Gruescu

Abstract Crawling mechanical structures with energetic autonomy and remote control may move and steer by sliding on the ground. The paper presents a study on simple possibilities to achieve sliding movement with a crawling tetrahedral structure and studies their movement capabilities both analytically and by MBS simulation. The structures are analyzed using an equivalent plane model and a 3d model, respectively. The analysis is structural, kinematical, static and dynamic.

Keywords Piezo-actuators • Hysteresis • Micro-mechanisms • Multi body system simulation

1 Introduction

Mankind has always manufactured means to displace material goods or persons. For a long time, the vehicles were driven by human or animal energy. The development of science and technology contributed to improvement of speed by means of different types of actuation (combustion engine, electric, hydraulic motor). Also, the reliability and safety increased as new materials, technologies and control devices became widely available. Still, the basic mechanical structures, such as wheels, remained the same.

Recently, historically speaking, people became aware of the fact that motion is a fundamental property of any living creature and started to think that the motion of mechanical devices might be improved by getting inspiration from biological structures. Studies of biomechanics were planned to investigate multiple aspects such as principles of motion, speed and energetic efficiency in relation with the number and size of the elements in the structure, symmetry in displacement, ratio of

D. Mărgineanu (✉) • E.-C. Lovasz • C.M. Gruescu
Department of Mechatronics, University Politehnica Timisoara, Timisoara, Romania
e-mail: dan.margineanu@upt.ro; erwin.lovasz@upt.ro; corina.gruescu@upt.ro

K.-H. Modler
Institut für Leichtbau und Kunststofftechnik, Technical University Dresden, Dresden, Germany
e-mail: karl-heinz.modler@tu-dresden.de

E.-C. Lovasz et al. (eds.), *Microactuators and Micromechanisms*, Mechanisms and Machine Science 30, DOI 10.1007/978-3-319-15862-4_2

segments' length and so on. Dickinson et al. offered an early general view on animals' movement (Dickinson et al. 2000).

Zielinska provided a review of all time walking machines and observations on displacement pattern of a large group of animals starting with simple invertebrates up to complex mammals (Zielinska 2004). Further extensive and substantial studies on imitation of human locomotion in robots' gait were presented by Zielinska et al. (Zielinska 2009; Zielinska et al. 2009; Zielinska and Chew 2006).

Nowadays engineering develops modules based on biological inspiration. The design of such structures is only at the beginning, but it spans quickly and intensively as people understood the significant potential of the new concept. Different parts of human or animal body are sources of inspiration for intelligent mechanical structures, which displace through sliding, crawling, climbing and stepping with a large variety of gaits.

Initially, only the human body was the inspiration for the so-called humanoid robots, a field with many successful results. So, a solution of human walking imitation in a robotic module was presented by Ceccarelli et al. (Liang et al. 2008; Li and Ceccarelli 2012). Ottaviano et al. designed and tested a hybrid leg-wheel walking machine (Ottaviano et al. 2011). Ceccarelli reviewed the progress in legged walking, emphasizing the importance of the kinematic and dynamic characteristics of the mechanical structure on the performance of the locomotion artifact (Ceccarelli 2013). The human body inspiration extended to other segments than legs. So, Liang and Ceccarelli proposed a robotic parallel structure meant to achieve the motion tasks performed by the human torso (Liang and Ceccarelli 2012).

More recently, other types of displacement were taken into account and studied in order to be imitated in robotic applications. One of the natural possibilities to displace is crawling or sliding-crawling, which became of intense interest. There are studies on fundamental physics involved in crawling structures. For instance, Wagner and Lauga presented a thorough insight on physics of locomotion taking into account mass, elastic connections, anisotropic friction, degrees of freedom, asymmetry of kinematics and friction forces (Wagner and Lauga 2013). In Ijspeert's work the biology and robotics meet to provide a crawling device based on the model of a salamander, both from mechanical and control point of view (Ijspeert 2008).

Quillin's researches focused on models of crawling taken from nature. His work provided thorough studies on kinematics of locomotion and various parameters which describe its efficiency in terms of speed and energy consumption. Furthermore, scaling of natural structures is an important issue to study as performances may vary significantly in respect with the size of the structure elements (Quillin 1999).

There are already practical results in crawling robots field. There are numerous solutions, more or less similar to natural displacement, which use various types of mechanisms such as parallelogram, pantograph and others. The resulting gait may be sliding, sliding-crawling as described by Tian et al. (2014). Zimmermann et al. explored the possibility to get inspiration from natural crawling movement of snakes or worms in the design of robots designated to operate in unpredictable environments. Their solution uses a magnetizable elastic body, which travels in a magnetic field (Zimmermann et al. 2007).

Liu et al. proposed a robotic earthworm with a biomimetic sensing system, working on a principle based on peristaltic contractions and elongations of worm's compliant segmented body (Liu et al. 2006).

Nam et al. developed a crawling micro robot that is able to move along a tube, such as a human blood vessel, by means of asymmetric friction force controlled with an oscillating external magnetic field (Nam et al. 2014). Wang et al. created a kinematic model, then a prototype of a mini-robot which crawls similarly to caterpillars (Wang et al. 2009).

The idea of designing crawling robots went beyond imitation of mechanical structure. Neuronal connections of different segments can also be mimicked, as stated by Lipson and Pollack (2000).

The conclusion of the brief presentation above is that crawling robots are a current subject of interest in engineering. The principles of locomotion are taken from certain species of animals, according to the skill considered to be most significant. The mechanical structures proposed until now are based on simple planar mechanisms, involving linkages and wheels. Thus, there is a broad development perspective in designing efficient and compact solutions.

The authors published previous results on crawling mechanisms in (Lovasz et al. 2005; Modler et al. 2005; Mărgineanu et al. 2007).

The proposed basic crawling mechanism is a pyramid with triangular basis (tetrahedron) with elastic joints and a pushing element linked on it's point. The movement obtained through sliding structure on the ground depends both on the friction in the contact points and ground surface planarity

The mechanism with one active element on an edge (Fig. 1a) or on the pushing element (Fig. 1b) touches the ground in 4 points, 3 of them on the basic link and the 4-th to the pushing element. In order for a sliding movement to appear when

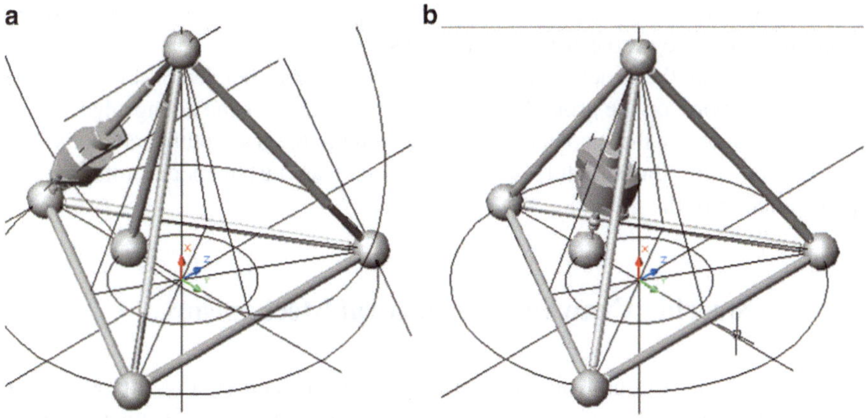

Fig. 1 Non-steering crawling tetrahedral mechanism (**a**) with active element on the lateral edge and (**b**) with active pushing element

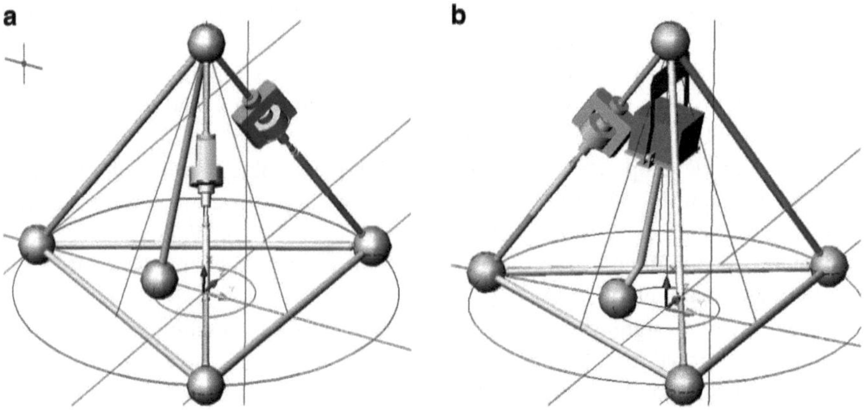

Fig. 2 Steering crawling tetrahedral mechanism (**a**) with sequentially actuators on the lateral edges and (**b**) with servomotor on the bended pushing element

the active element is actuated, the friction force between the pushing element and the ground has to be greater then the sum of all friction forces in the basis link's contact points. This mechanism cannot steer its movement.

In the Fig. 2, two steering possibilities with additional actuators are shown. Steering supposes the control of the position and rotation vectors, so the mobile's characteristic point (CP) should move along the desired path and the characteristic axis (CA) remains tangent to the path. If for some reason the CP leaves the path or the CA twists, the steering should compensate by imposing a counter sideways movement or a counter twist. The steering can be done by lateral or oblique pushing, by altering the reactions on the bases touch points or by both methods.

As the two actuators on the front lateral edges (Fig. 2a) move at the same time, the structure glides as a non-steered one. When only one side acts, the passive pushing element presses obliquely on the ground relative to a CA rotated in the direction of the leading edge and the lateral component (in z axis direction) of the frictional force moves the structure sideways.

In the Fig. 2b, the structure uses a servomotor to turn a bended pushing element, so the touch point on the ground is no longer on the CA. As the acting edge shrinks the reactions on the three contact points between the basis structure and the ground are uneven, thus the structure will twist around an instantaneous rotation axis perpendicular to the ground.

2 Non-steering Crawling Tetrahedral Mechanisms

In the following, the non-steering tetrahedral mechanism (s. Fig. 1a) with rigid basis contour and actuated edge element will be analyzed by an equivalent plane structure (s. Fig. 3). This equivalent plane mechanism with revolute joints instead of the compliant joints used has the degree of freedom:

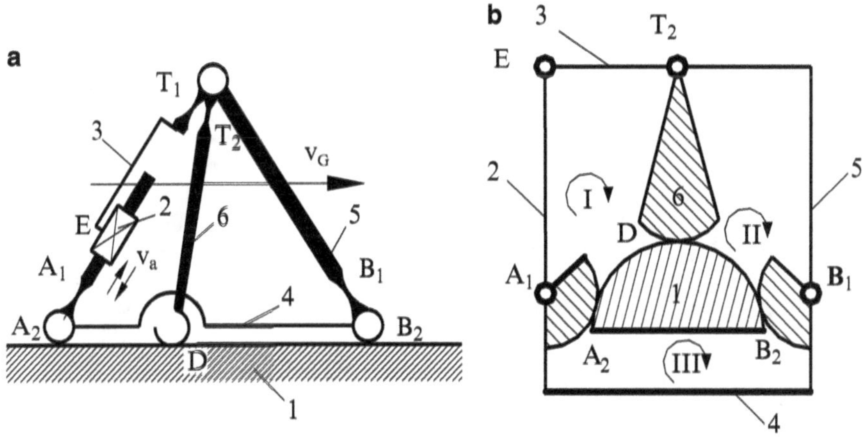

Fig. 3 Equivalent plane structures (**a**) and structural schema of non-steering crawling mechanism (**b**)

$$F = 3 \cdot (n - 1) - 2 \cdot e_1 - e_2 = 3 \cdot 5 - 2 \cdot 5 - 1 \cdot 3 = 2, \qquad (1)$$

where n is the number of elements and e_1 and e_2 the joints number with 1 and 2 degrees of freedom, respectively (s. Fig. 3a).

The crawling mechanism realizes a constraint movement as the dynamical constraint movement condition is fulfilled, i.e.:

$$n_a < F \leq n_a + n_{din} = 2 \qquad (2)$$

where: n_a is the number of actuating elements and n_{din} the number of elements, which realizes the dynamic equilibrium.

For slow actuation the dynamic forces may be neglected and the sliding conditions for the mechanism are (see Fig. 4):

$$F_{D16} > F_{A14} + F_{B14} \Rightarrow \mu \cdot G_{D16y} > \mu \cdot \left(G_{A14y} + G_{B14y}\right). \qquad (3)$$

The joint reactions $G_{A,B14y}$ are determined by the kinetostatic analysis (presented in (Modler et al. 2005; Mărgineanu et al. 2007)).

The mechanism does not lose the contact with the ground in A and B if:

$$G_{A14y} > 0, \;\; G_{B14y} > 0. \qquad (4)$$

The tetrahedron's weight vector has to pass between the points K_A and K_B, i.e. the corresponding intersections between the reaction's direction in the joints A and B and the reaction's G_{65} direction on the pushing element.

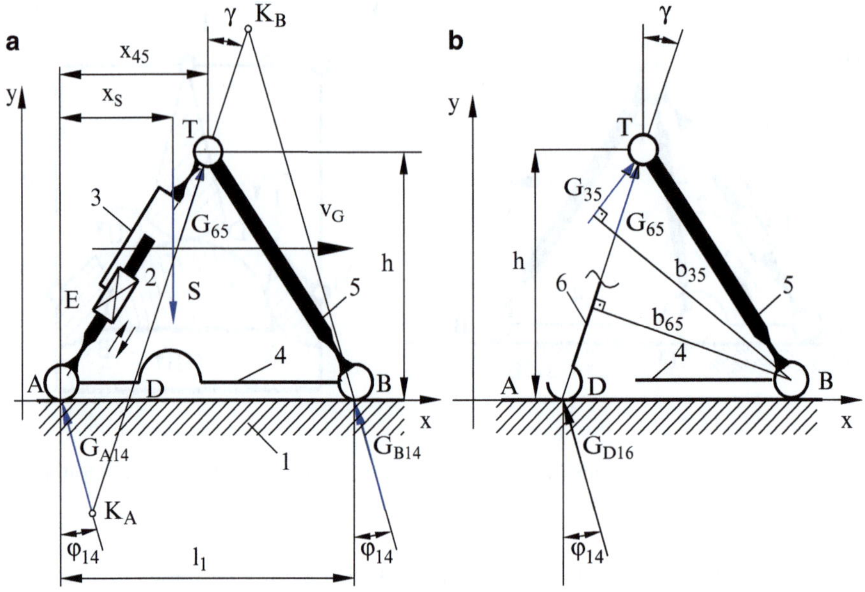

Fig. 4 Kinetostatics of the non-steering crawling mechanical structure

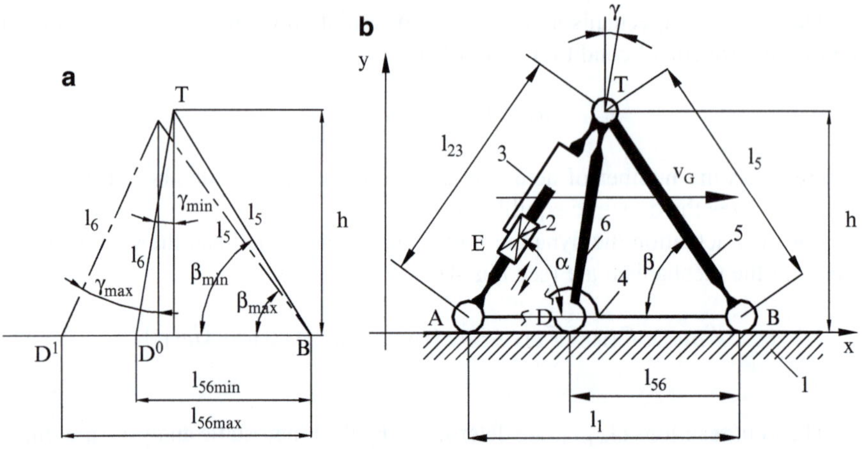

Fig. 5 Kinematic of the non-steering crawling tetrahedral mechanism - geometry (**a**) and structure (**b**) aspects

The sliding speed depends on actuator's velocity v_a and the angles α and γ (Fig. 5):

$$v_G = v_T = v_a \cdot \cos\alpha - l_{23} \cdot \frac{d\alpha}{dt} \sin\alpha - l_6 \cdot \frac{d\gamma}{dt} \cos\gamma. \tag{5}$$

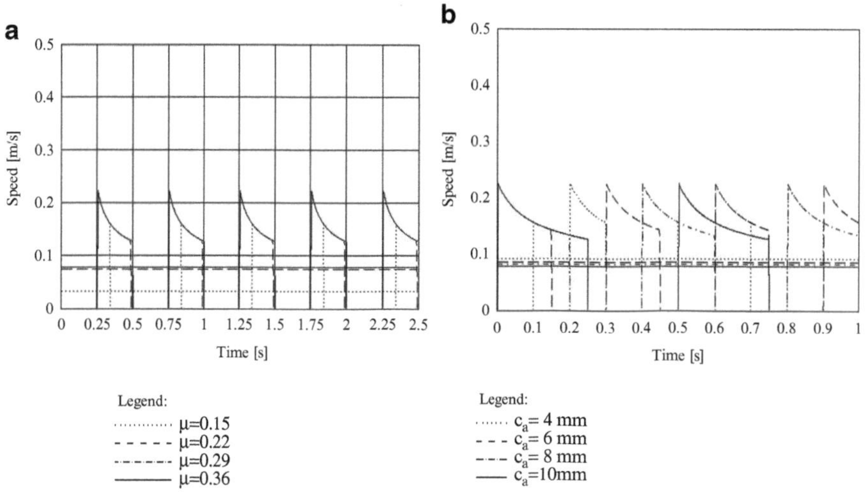

Fig. 6 The influence of frictional coefficient (**a**) and actuator's stroke (**b**) on mechanism's speed

The push does not slide on the ground if $\gamma < \phi_{16}$.

In (Modler et al. 2005), the dependency of the instant and average sliding speed (Fig. 6a, b), on the frictional coefficient and actuator's stroke was studied.

As to obtain more accurate results for the movement capabilities of the crawling tetrahedral mechanisms, the inertia of the elements is not to be neglected. The difference between static and dynamic friction is also important because of the intermittent sliding of the mechanism. As analytical calculus leads to a non-linear differential equations system, the RBS method for dynamical analysis was chosen. A CAD model for the mechanism wad built using SolidWorks, as presented in Fig. 7.

The model uses solid bodies and revolute joints to simulate the compliant joints of the mechanism an one driving prismatic joint to simulate the actuator.

The dynamic analysis was done by Rigid Body Simulation using Motion Analysis in SolidWorks Motion. So, the inertia and different friction conditions between the elements and the ground may be taken into account, including different values for the static and dynamic friction coefficient, important for studying a crawling motion with low speed and stop/slide intervals.

With this simulation model, the influence of actuation stroke and frequency on the movement capabilities of the tetrahedral mechanism may be studied, as well as the contact reactions of the ground, the loads in the joints and the energy consumption.

In Fig. 8, simulation results for the model non-steering crawling tetrahedral mechanism are presented. The simulated path (Fig. 8a) is linear, with some instability at the beginning of the simulation produced by the phenomena associated to the contact conditions initiation. The sliding speed (Fig. 8b) does not have the sharp peaks as in Fig. 6 anymore and some instability at the beginning of the motion is also to be detected, due to inertia and friction.

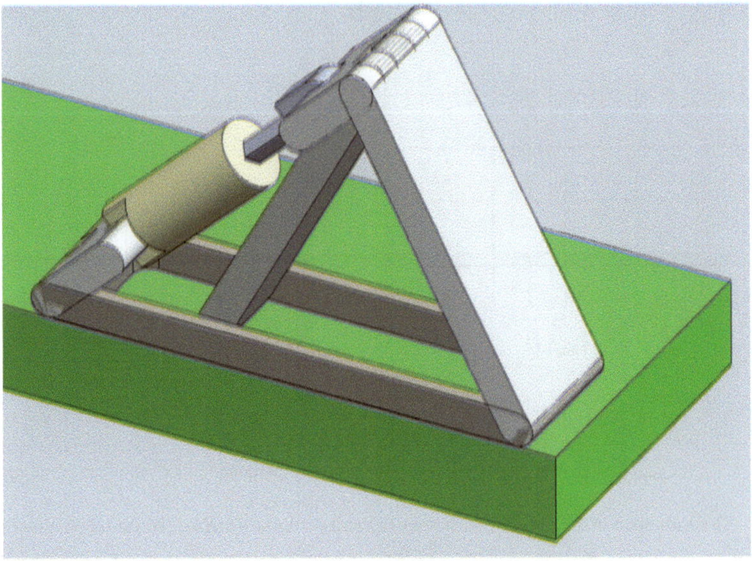

Fig. 7 CAD model of the non-steering crawling tetrahedral mechanism

3 Steering Crawling Tetrahedral Mechanism

The steering crawling mechanism with solid base link, one actuated edge and rotating pushing element (s. Fig. 2b) is composed of seven 3D elements, with 6 degrees of freedom each. These elements are linked by compliant joints 4 joints (E, F, T_1, P_2–P_3) with equivalent 1 degree of freedom and by mechanical joints as follows: 1 joint (T_2) with 2 degrees of freedom, 1 (P_1) with 3 degrees of freedom and 4 joints (A, B, C, D) with 5 degrees of freedom:

The equivalent mechanism (s. Fig. 9) has the degree of freedom:

$$F=6\cdot(n-1)-\sum_{i=1}^{5}(6-f_i)\cdot e_i=6\cdot 6-(5\cdot 4+4\cdot 1+3\cdot 1+1\cdot 4)=5. \quad (6)$$

The structure has a constraint motion because:

$$n_a < F \leq n_a + n_{din} = 5, \quad (7)$$

where n_a is the number of acting movements and n_m is the number independent movements: slide, drift and twist, determined by the kinetostatical conditions.

The movement conditions for the mechanism with solid base, one active edge element and turning passive bent push was be studied in (Mărgineanu et al. 2007) order to generate sliding movement and steer it on the desired path. The mechanism's movement depends on the two acting elements, i.e. the length l_{34} of the

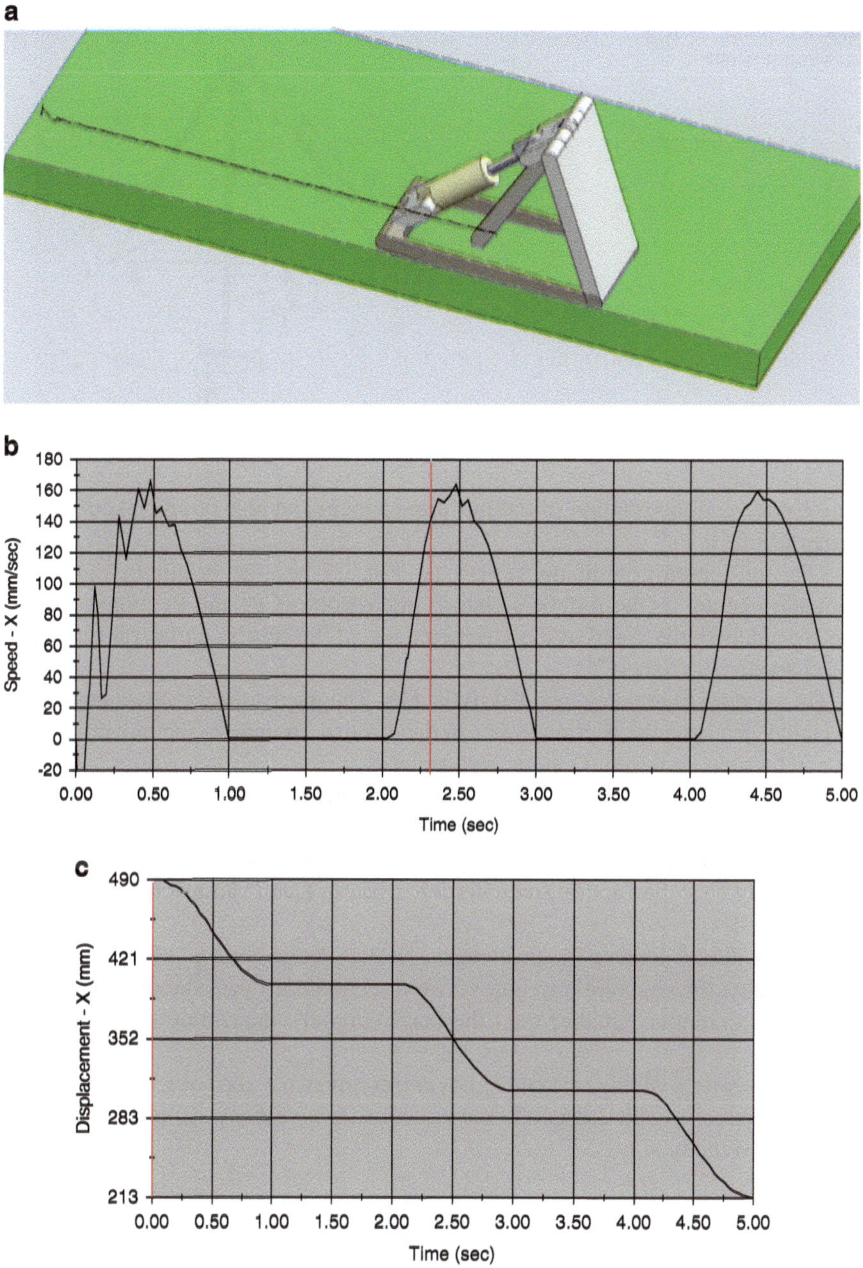

Fig. 8 Simulation results for the model of non-steering crawling tetrahedral mechanism: (**a**) the simulated linear path, (**b**) the crawling velocity variation and (**c**) the linear displacement of the structure

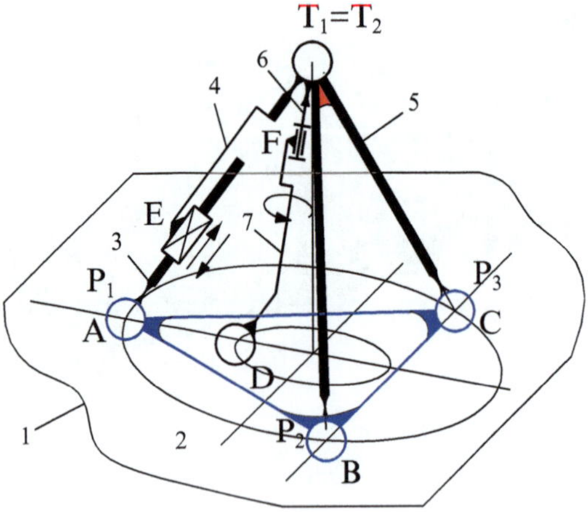

Fig. 9 Equivalent 3D structure for the steering crawling structure

active edge and the rotation of the pushing element and also on the position of its center of mass.

The considered coordinate system (s. Fig. 2) has the origin in the base's symmetry center as the mobile's characteristic point (CP) and the plane xOy on the ground, with the x-axis as the symmetry axis of the base's equilateral triangle as the characteristic axis (CA).

The pushing direction is parallel to the CA. The mechanism's movement on the ground depends on the friction forces in the 4 contact points, so the movement will be conditioned its weight G value and center

On the structure act the resultant joint reactions R_j in the three contact points between the basis contour and the ground and R – in the tip of the pushing element, with the components N_j, respectively N normal to the contact surface and F_{fj}, respectively F_f – the friction forces in these points (see (Mărgineanu et al. 2007) and Fig. 10).

The conditions to keep the mechanism on the ground are $N_j > 0$.

Generally, the structure's weight vector direction must pass through the interior of a certain contour (smaller than the basis contour), depending on the pushing angle.

The structure's sliding movement has two components: the velocity v_O of the CP on the trajectory, which depends on the actuator's velocity v_a, and the angular (twisting) velocity ω_O:

$$\omega_O = \frac{v_O}{\rho_O}, \tag{8}$$

where $\rho_O = \sqrt{{x_M}^2 + {y_M}^2}$ is the distance between the instantaneous rotation point and the structures CP.

The CAD model for the 3D steering crawling mechanism presented in Fig. 11 has a push that can be positioned along the upper edge, so the pushing on the ground

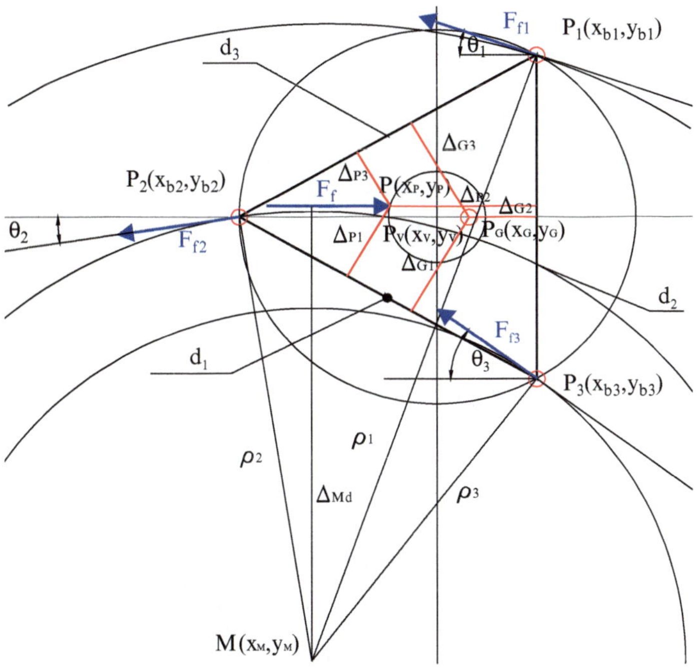

Fig. 10 Kinetostatic of the steering crawling mechanical structure

is asymmetrical. The path of the mechanism shows an obvious curvature, as the pressure distribution on the ground is not even and the friction twists the mechanism. By adjusting the position of the push, different curvatures may be obtained (Fig. 12).

Steering and non-steering mechanisms' physical models built and tested both in TU Dresden and UP Timişoara by joint teams also include power batteries to achieve some moving independence and carry all the electronics and sensors needed to act and control their movement (Fig. 13).

4 Conclusions

The movement capability of the tetrahedral crawling mechanisms is influenced by the mechanism's geometry and the friction conditions. The dependence of the steering on the position of the gravity center, on the planarity of the ground and especially on the friction conditions in the 4 contact points with the ground shows, as experiments confirmed, that controlling the movement only by programmed actuators is not precise. In order to increase the steering accuracy, feedback informations about the CP's position and the CA's direction must be collected

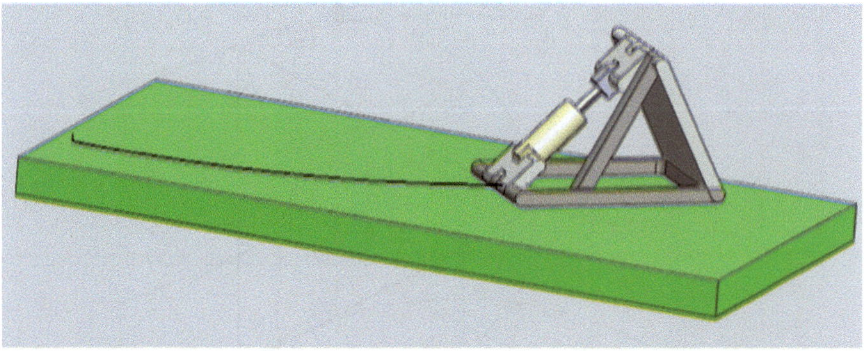

Fig. 11 3D CAD model and simulation result for the steering crawling tetrahedral mechanism

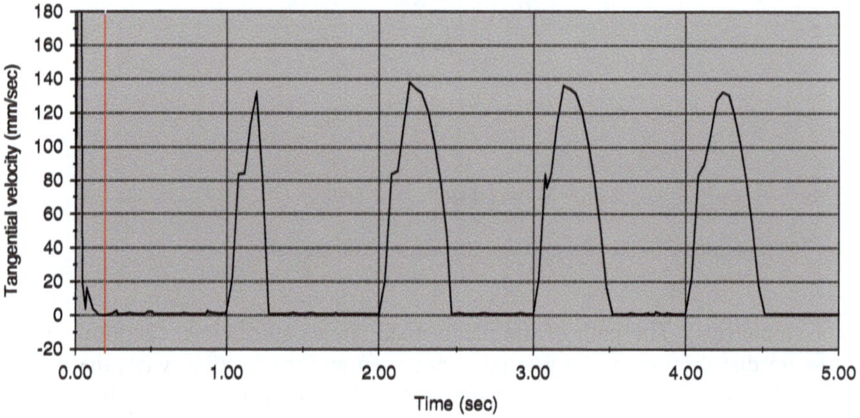

Fig. 12 Simulation result for the steering crawling tetrahedral mechanism: the tangential velocity of the mechanism

from the environment and the actuators movement adjusted so the structure stays on track.

Modeling and simulations allow a more accurate estimation of the steering capabilities and the obtainable speed of the mechanism in the design stage.

Simulations with high speed and low stroke actuators have shown that the movement may be more stable, as the inertia does not allow the sliding to stop in the retracting phase of the actuator.

At even higher speed, the push is no longer needed, as the mechanism acts like a vibrating plate on the ground. Dynamic studies for these cases will be published by the authors in further works.

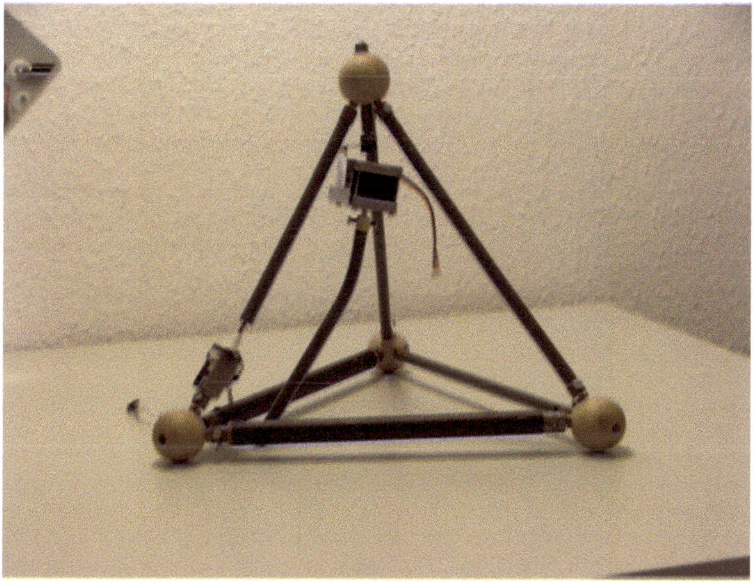

Fig. 13 The steering crawling tetrahedral mechanism

References

Ceccarelli M (2013) Mechanism design for robots. In: The 11th IFToMM international symposium on science of mechanisms and machines SYROM'13, Brasov. Springer, Dordrecht, pp 1–8

Dickinson MH et al (2000) How animals move: an integrative view. Science 288:100

Ijspeert AJ (2008) Central pattern generators for locomotion control in animals and robots: a review. Neural Netw 21:642–653

Li T, Ceccarelli M (2012) A method for topological design of mechanism. In: Proceedings of the MEDER 2012, 2nd IFToMM symposium on mechanism design for robotics, Beihang University, Beijing

Liang C, Ceccarelli M (2012) Design and simulation of a waist–trunk system for a humanoid robot. Mech Mach Theory 53:50–65

Liang C, Ceccarelli M, Takeda Y (2008) Operation analysis of a one-DOF pantograph leg mechanisms. Proceedings of the RAAD, the 17th international workshop on robotics in Alpe–Adria–Danube Region, Ancona, pp 1–10

Lipson H, Pollack BJ (2000) Automatic design and manufacture of robotic life forms. Nature 406:974–978

Liu W, Menciassi A, Scapellato S, Dario P, Chen Y (2006) A biomimetic sensor for a crawling minirobot. Robot Auton Syst 54:513–528

Lovasz EC, Modler KH, Cărăbaş I (2005) Internationale Zusammenarbeit zwischen PU Timişoara und TU Dresden auf dem Gebiet der Getriebelehre – Getriebekolloquium 2005, RWTH Aachen, pp 179–192

Mărgineanu D, Lovasz EC, Modler KH (2007) On the 3D crawling mechanical structures. The 12th IFToMM world congress, Besançon

Modler KH, Mărgineanu D, Perju D, Lovasz EC, Fernengel V (2005) Analyse mechanischer Strukturen für einfache Fortbewegungen. In: Proceedings of the ninth IFToMM international symposium on the theory of machines and mechanisms, vol 1. Bucharest, pp 105–110
Nam J, Jeon S, Kim S, Jang G (2014) Crawling microrobot actuated by a magnetic navigation system intubular environments. Sens Actuat A-Phys 209:100–106
Ottaviano E, Vorotnikov S, Ceccarelli M, Kurenev P (2011) Design improvements and control of a hybrid walking robot. Robot Auton Syst 59:128–141
Quillin KJ (1999) Kinematic scaling of locomotion by hydrostatic animals: ontogeny of peristaltic crawling by the earthworm lumbricus terrestris. J Exp Biol 202:661–674
Tian Y, Wei X, Joneja A, Yao YA (2014) Sliding–crawling parallelogram mechanism. Mech Mach Theory 78:201–228
Wagner GL, Lauga E (2013) Crawling scallop: friction-based locomotion with one degree of freedom. J Theor Biol 324:42–51
Wang W, Wang K, Zhang H (2009) Crawling gait realization of the mini-modular climbing caterpillar robot. Prog Nat Sci 19:1821–1829
Zielinska T (2004) Biological aspects of locomotion. In: Pfeiffer F, Zielinska T (eds) Walking: biological and technological aspects, vol 467, CISM courses and lectures. Springer, Wein/New York, pp 1–30
Zielinska T (2009) Biological inspiration used for robots motion synthesis. J Physiol Paris 103:133–140
Zielinska T, Chew CM (2006) Biologically inspired motion planning in robotics. In: Kozlowzki K (ed) Robot motion and control, vol 335, Lecture notes in control and information sciences. Springer, Heidelberg, pp 201–219, Chapter 13
Zielinska T, Chew CM, Kryczka P, Jargilo T (2009) Robot gait synthesis using the scheme of human motion skills development. Mech Mach Theory 44(3):541–558
Zimmermann K, Naletova VA, Zeidis I, Turkov VA, Kolev E, Lukashevich MV, Stepanov GV (2007) A deformable magnetizable worm in a magnetic field—a prototype of a mobile crawling robot. J Magn Magn Mater 311:450–453

New Kinematic Designs of Flexure Hinge Based 3 DoF Translational Micromanipulators

I. Prause, D. Schoenen, and B. Corves

Abstract The development of new robot structures, in particular of parallel kinematic structures, is widely systematized by different structure synthesis methods, whereas a systematic approach for the development of micromanipulators is missing. Hence the aim of this paper is to systematically develop new kinematic designs of flexure hinge based micromanipulators with three translational degrees of freedom. With the focus on screw-based approaches for the structural synthesis a structural selection is made with regard to specific characteristics of micromanipulators. Thus 11 possible configurations of new kinematic structures with a symmetrical and orthogonal assembly are shown. Finally, an internal development of a flexure hinge based micromanipulator is presented.

Keywords Micromanipulators • Compliance • Flexure hinge • Structure selection • Kinematic design

1 Introduction

Micromanipulators with parallel kinematic structures have a substantial payload ratio. Furthermore, the actuators can be mounted on the frame which reduces the inertia of moving masses and increases the dynamic performance (Merlet 2006). Thus, parallel micromanipulators have potential for increasing precision. This is in line with the general technological development, which has led to a decrease in object dimensions but an increase in need for precision, for example, in IT-technology, bio-medicine or fiber technology.

The recent growth in compliant mechanisms applications is due to advances in understanding of compliant mechanisms and many benefits on the one hand like the

I. Prause (✉) • D. Schoenen • B. Corves
Institut für Getriebetechnik und Maschinendynamik, RWTH Aachen University, Aachen, Germany
e-mail: prause@igm.rwth-aachen.de; schoenen@igm.rwth-aachen.de; corves@igm.rwth-aachen.de

E.-C. Lovasz et al. (eds.), *Microactuators and Micromechanisms*, Mechanisms and Machine Science 30, DOI 10.1007/978-3-319-15862-4_3

monolithic structure, which leads to reduced assembly effort, reduced wear, reduced need for lubrication and weight reduction (Howell et al. 2013). Furthermore, a monolithic structure also results in close to zero backlash, frictionless motion, significant potential for miniaturization and increased performance such as high precision (Howell 2001). On the other hand, the disadvantages of compliant mechanisms are, for example, undesired parasitic effects, deviation in motion, difficulties in analyzing and designing and a limited range of motion (Howell 2001). But for high precision parallel micromanipulators with a workspace typically in the millimeter range and macroscopic dimensions the mentioned disadvantages are no drawbacks. Besides, for most applications of micromanipulators only translational motions are necessary.

Furthermore, the current literature does not provide a systematical approach for new developments of micromanipulators. Rather, it seems that recent developments are initiated by means of heuristic approaches. A comparison and benchmark of different structures is missing as well.

In a classification scheme of robots is already provided and several existing micromanipulators are compared. Thus the main focus of this paper is the systematical development of new kinematic designs for flexure hinge based parallel micromanipulators with three translational degrees of freedom (DoF).

In Sect. 2 the considered joints in this study are described. Furthermore, a systematic approach for the design of new micromanipulators is introduced, followed by an appropriate selection scheme. Thereafter, in Sect. 3, possible new kinematic structures with three translational DoF for micromanipulators are shown. A flexure hinge based micromanipulator developed at our department is presented in Sect. 4. Finally, Sect. 5 gives a brief conclusion of the paper and a perspective on further research projects.

2 Further Developments Towards New Structures

In the following chapter new structures for micromanipulators are developed. These structures should meet different requirements: (1) 3 DoF translational motion, (2) parallel kinematic structure, (3) if possible, symmetrical assembly.

Firstly, the considered joints are introduced and described (Sect. 2.1). Thereafter, possible joint configurations for the limbs are presented followed by an appropriate selection scheme for the design of new structures (Sect. 2.2).

2.1 Considered Joints

Various compliant joints are extensively described in the literature by Raatz (2006) and Howell et al. (2013). Nevertheless, due to compact size and simplicity, flexure

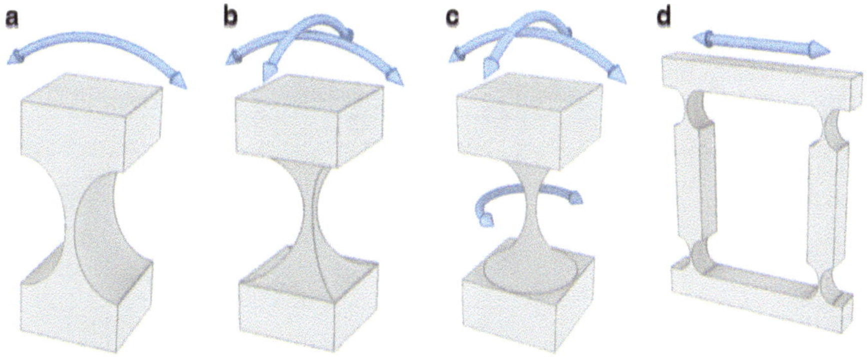

Fig. 1 Considered joints for the further development. (**a**) Revolute joint (R). (**b**) Universal joint (U). (**c**) Spherical joint (S). (**d**) Prismatic joint (P) (based on four R's)

hinges with lumped compliance are the most common. To allow flexibility in one or more directions the hinge blank is notched. This enables different DoF. According to Fig. 1 (a) revolute joints are manufactured by necking down the hinge blank once (b) universal joints by cutting it twice or (c) for spherical joints by turning the hinge blank while notching. Revolute joints are easy to cut by wire-EDM for planar and spatial applications. Thus, in compliant mechanisms universal and spherical joints are often substituted by a combination of two or three revolute joints. In monolithical structures prismatic joints have to be replaced by a parallelogram structure, based on four revolute joints (cf. Fig. 1). Hence within this paper only structures with revolute joints or prismatic joints substituted by a parallelogram are considered.

2.2 *Description of Possible Configurations and Selection Scheme*

The selection scheme for kinematic structures offering new possibilities for micro-manipulators is based on the constraint-synthesis approach, especially the virtual chain approach introduced by (Kong and Gosselin 2007). This approach uses the screw theory to synthesize new kinematic structures for parallel manipulators with three identical limb configurations. To ensure a fully translational motion of the end-effector, the joint alignment cannot be chosen arbitrarily. The *twist*-system of the manipulator is a 3-ξ_∞-system (translational motion, screws with infinite pitch), the *wrench*-system is a 3-ζ_∞-system respectively (moments around the origin of the reference frame) because the virtual power developed by any ζ_∞ along any twist within the 3-ξ_∞-system is zero. The *wrench*-system of the manipulator is a linear

Table 1 Structure selection

Class	No. of configurations (NoC)	NoC after applying criteria 1 + 2	NoC after applying criteria 3 + 4	Resulting limb configurations
3P	1	–	–	–
3R-1P	4	4	1	$\underline{P}R_1R_1R_1$
2R-2P	6	6	3	$\underline{P}R_1R_1P$, $\underline{P}R_1PR_1$, $\underline{P}PR_1R_1$
1R-3P	4	–	–	–
5R	5	5	–	–
4R-1P	30	10	2	$\underline{P}R_1R_1R_2R_2$, $\underline{P}R_1R_2R_2R_1$
3R-2P	30	–	–	–
2R-3P	10	–	–	–

Based on Kong and Gosselin (2007)

combination of the *wrench*-systems of the limbs, whereas the *wrench*-system of limb *i* is the intersection of the *wrench* systems of all the joints in the limb kinematic chain. At least one rotation of the end-effector should be blocked by one limb.

Only simple joints such as prismatic (P) and revolute joints (R) are considered. Restricting the maximum number of joints within one limb to five implies eight different classes (cf. Table 1). For each class several joint alignments are possible, so that the number of configurations within each class can be up to 30. A more detailed description of the possible combinations gives (Kong and Gosselin 2007).

To reduce the number of configurations (NoC) – from 90 in total – the following criteria are applied: (1) within one limb the number of prismatic joints should not exceed two, (2) no inactive joints, (3) first joint of each limb is actuated, (4) actuated joints are only prismatic joints. The last criterion is assumed because it is not possible to actuate a rotational axis of a flexure hinge directly in an appropriate accuracy range. Table 1 gives an overview of the resulting NoC after applying the above mentioned criteria.

An underlined letter indicates an actuated joint, whereas the index describes the joint alignment within one limb. An identical index means that the joint axes are oriented parallel to one another.

For the resulting limb configurations an orthogonal assembly is desired. The reasons are as follows: An orthogonal assembly of the entire structure implies actuation axes that are perpendicular to each other. It also enables an orthogonal limb structure, which means that the different rotation axes within one limb can be arranged perpendicular to each other. This arrangement simplifies the manufacturing of limbs cut by advanced wire-EDM technology because of reduced clamping effort. Thus, in return form deviations are reduced and the positioning accuracy of the entire structure increases. Orthogonal limb design also supports compact over all structures as far as the different limbs can be nested in one another.

Table 2 Restrictions for orthogonal assembly

No.	Configuration	Direction of prismatic joint axis (actuated) within a limb (Gogu 2009)	Direction of prismatic joint axis (non-actuated) within a limb (Gogu 2009)
1	$\underline{P}R_1R_1R_1$ *)	$P \parallel R_1$	–
2, 3	$\underline{P}R_1R_1P_a$	(1) $P \perp R_1$ and (2) $P \parallel R_1$	(1) $P_a \parallel R_1$ and (2) $P_a \perp R_1$
4	$\underline{P}R_1P_aR_1$	$P \perp R_1$	$P_a \parallel R_1$
5, 6	$\underline{P}P_aR_1R_1$	(1) $P \perp$ R axes of P_a and	(1 + 2) $P_a \parallel R_1$
		(2) $P \parallel$ R axes of P_a	
7, 8	$\underline{P}R_1R_1R_2R_2$	(1) $P \perp R_1$ and (2) $P \parallel R_1$	–
9, 10, 11	$\underline{P}R_1R_2R_2R_1$	(1 + 2) $P \perp R_1$ and (3) $P \parallel R_1$	–

*)Additional assembly restriction (Kong and Gosselin 2007): all rotational joint axes are not parallel to a plane

Table 2 summarizes the different possibilities of joint alignments to achieve an orthogonal assembly based on (Gogu 2009). The motion axis of the actuated prismatic joint can be either parallel or perpendicular to the first rotational joint axis. For the configurations No. 2-6 with two prismatic joints, an additional criterion for the second prismatic joint is added. Due to the fact that in compliant mechanisms a prismatic joint is replaced by a parallelogram structure based on four revolute joints (cf. Sect. 2.1), the second prismatic joint within the limbs is labelled as P_a.

The proposed selection scheme and the assembly restrictions lead to 11 structures for translational micromanipulators with three DoF which can be further analyzed.

3 New Possible Structures

To give an impression of the derived limb configurations and the resulting parallel kinematic structures, they are presented in Figs. 2 and 4 (Annex) in numerical order according to Table 2. It is obvious that only revolute joints and parallelograms based on four revolute joints as prismatic joints are used. The limbs are each identically designed and orthogonally assembled. But to reduce the overall size of the entire structure sometimes it is necessary to rearrange the actuated parallelogram within the limbs. The actuated prismatic joints (attached to the fixed frame) are marked by an arrow. Besides, in Fig. 2, the kinematic structure of one limb is sketched in.

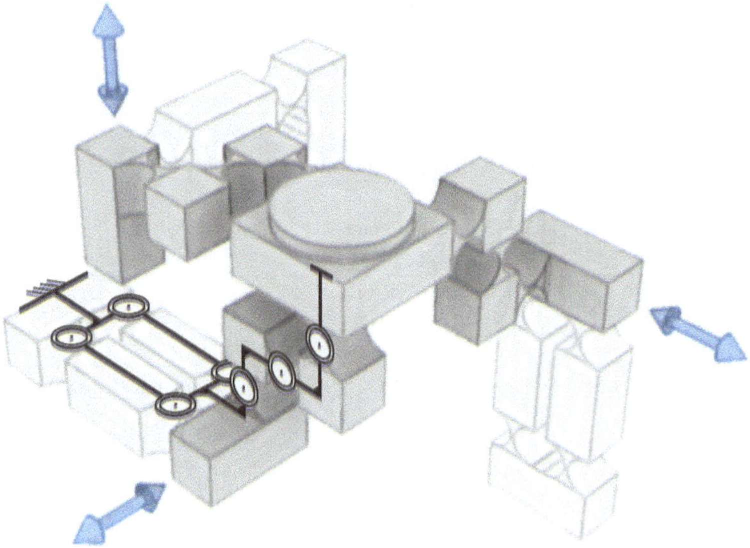

Fig. 2 $\underline{P}R_1R_1R_1$ – configuration with exemplary kinematic chain of one limb

4 Micromanipulator

Based on the previously derived structures a micromanipulator with three translational DoF is developed. A first draft of the micromanipulator is already given in (Ivanov and Corves 2012). Within the mentioned paper different possible limbs are considered, the $\underline{P}R_1PR_1$ – configuration is chosen and a stiffness analysis is given.

Based on these results the MicMaC (**Mic**ro **Ma**nipulator **C**ube) is built as shown in Fig. 3. The limb design is split into two parts. The actuated parallelogram $\underline{P}$ is separated from the contiguous components (R_1PR_1). Thus the two parts can be separately manufactured by wire-EDM even though the entire limb structure is not free of undercuts. To reach compact dimension of $180 \times 180 \times 180$ mm^3 the actuators are nested inside the parallelogram $\underline{P}$ which is attached to the base. The manipulator is driven by three Piezo LEGS® linear motors (resolution <1 nm). The motor mounting also holds the linear encoder system from Renishaw (resolution 0.1 µm). All flexure hinges are circular notch joints. To avoid excessive stress in the joints the working space of the micromanipulator is restricted to $\pm 2 \times \pm 2 \times \pm 2$ mm^3.

Due to the accuracy of the used components the repeatability can be assumed being in the submicrometre range. Nevertheless, the actual accuracy, as well as parasitic deviations, remains to be investigated in further research.

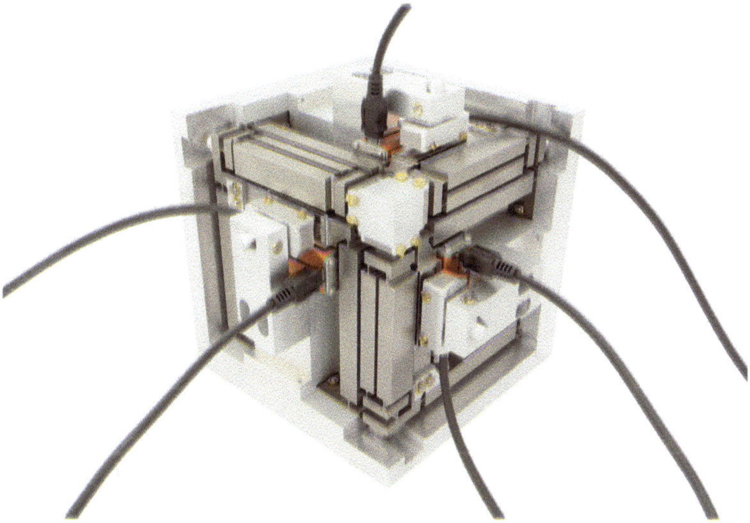

Fig. 3 MicMaC with $\underline{P}R_1P_aR_1$ – configuration

5 Conclusions

In this paper a method for the systematic development of flexure hinge based micromanipulators based on the constraint-synthesis approach was presented. Firstly, the utilized joints and their manufacturing process were presented. Then possible arrangements of the joints within a limb on the basis of several criteria were shown. By assembling these limbs in different ways, 11 orthogonal structures with a symmetrical architecture could be generated. Finally, based on a shown limb configuration the high precision MicMaC micromanipulator developed at our department was introduced.

Further developments should include the analysis of the developed structures to enable a comprehensive comparison and hereupon to make a selection for a final design. For this purpose different calculation algorithms (e.g. inverse kinematics, Jacobian matrix) have to be derived and implemented. In addition, a detailed consideration of stiffness properties in order to evaluate the positioning accuracy is to be made. Also the actual accuracy of the presented high precision MicMaC micromanipulator has to be validated and parasitic deviations have to be determined.

Acknowledgments The research work reported here was possible within a joint research project, which also includes the Laboratory for Machine Tools and Production Engineering and the Facility for Electron Microscopy of Aachen University and is funded by the German Grant Authority DFG.

Annex

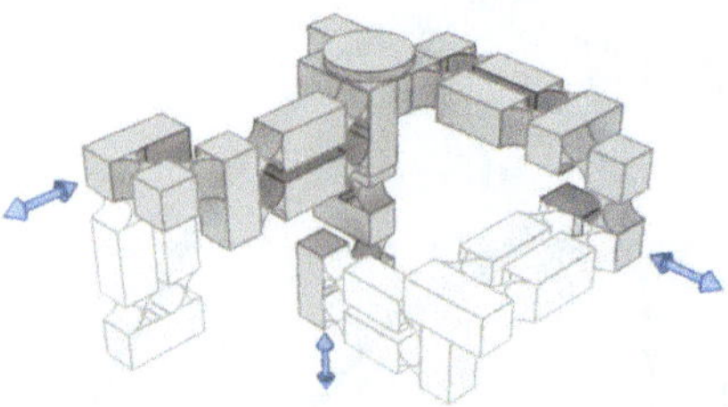

No. 2) $\underline{P}R_1R_1P_a$ - configuration

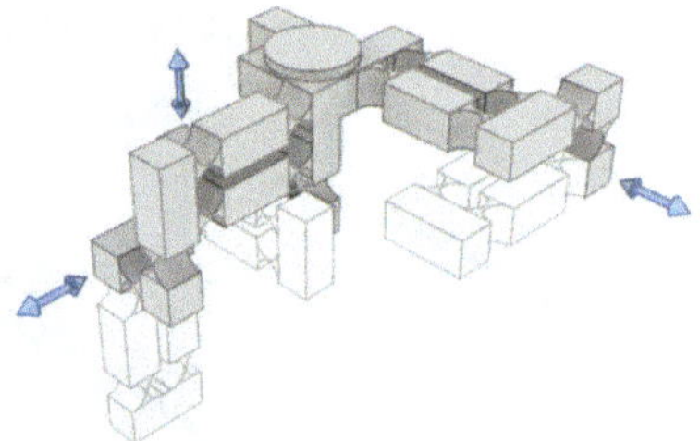

No. 3) $\underline{P}R_1R_1P_a$ - configuration

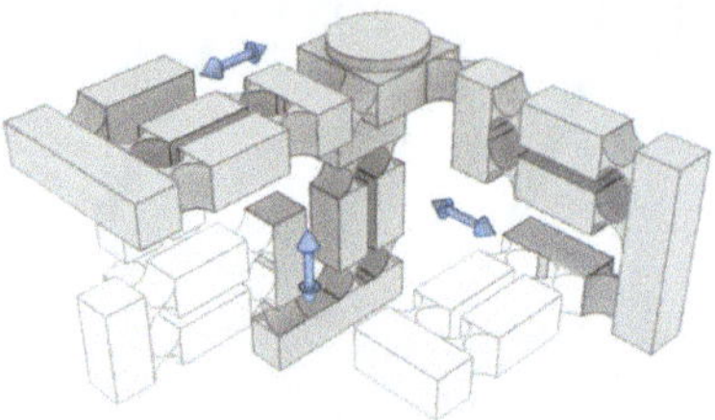

No. 4) $\underline{P}R_1P_aR_1$ - configuration

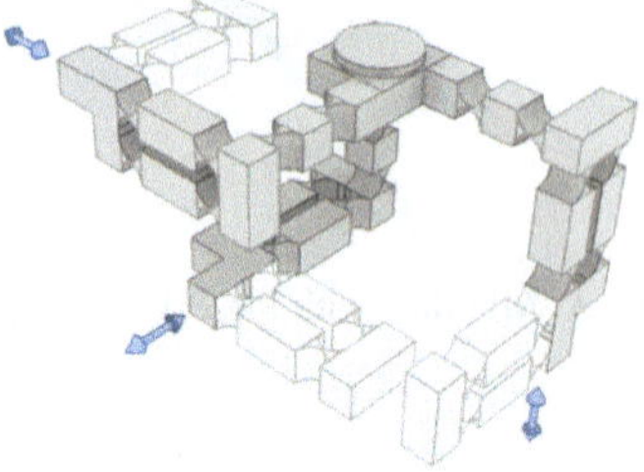

No. 5) $\underline{P}P_aR_1R_1$ - configuration

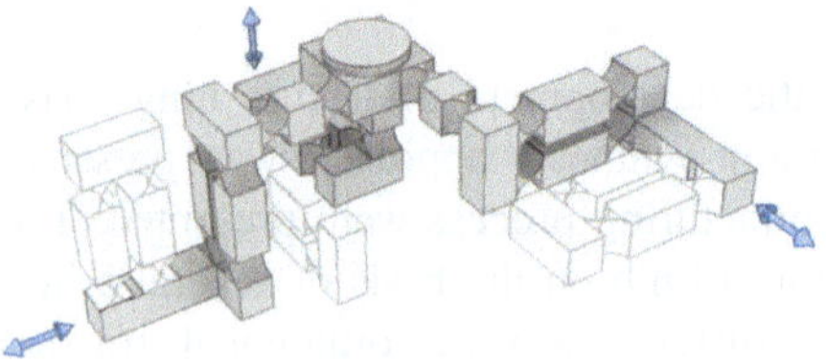

No. 6) $\underline{P}P_aR_1R_1$ - configuration

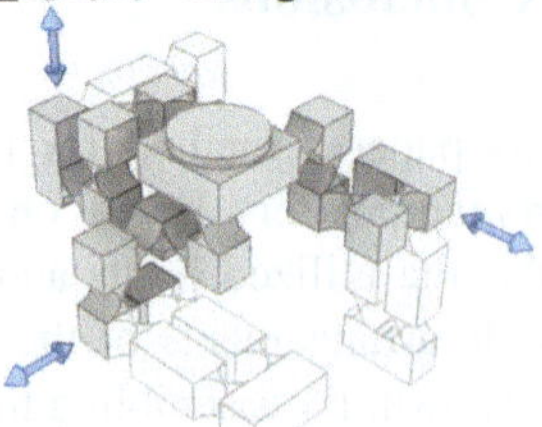

No. 7) $\underline{P}R_1R_1R_2R_2$ - configuration

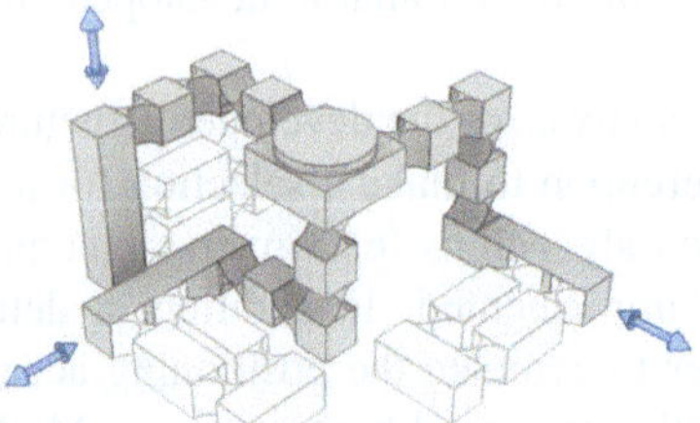

No. 8) $\underline{P}R_1R_1R_2R_2$ - configuration

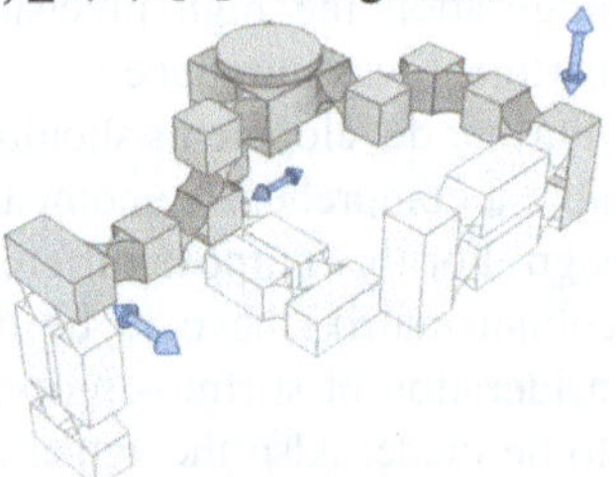

No. 9) $\underline{P}R_1R_2R_2R_1$ - configuration

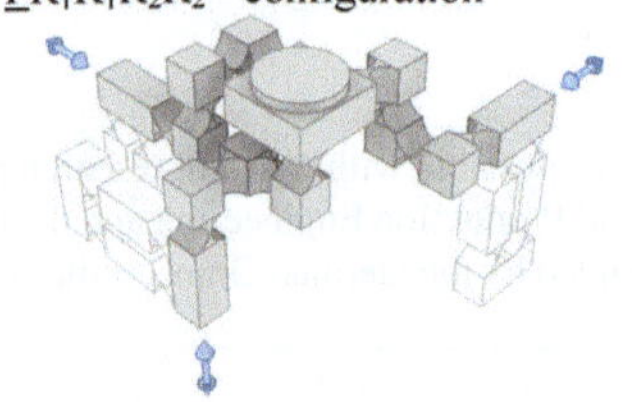

No. 10) $\underline{P}R_1R_2R_2R_1$ - configuration

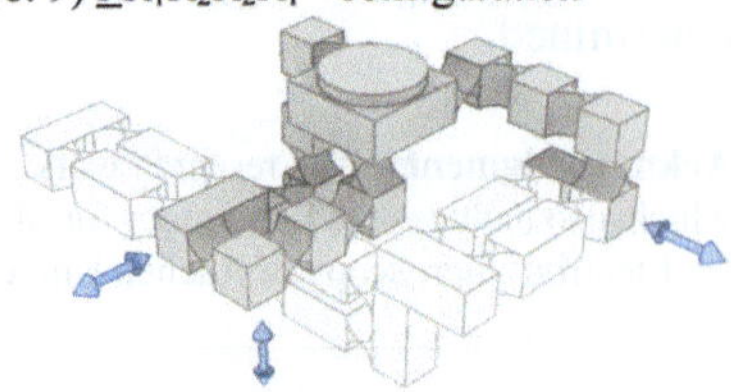

No. 11) $\underline{P}R_1R_2R_2R_1$ - configuration

Fig. 4 Schematic models of the different configuration according to Table 2

References

Gogu G (2009) Structural synthesis of parallel robots. Translational topologies with two and three degrees of freedom, vol 159, Solid mechanics and its applications. Springer, Berlin

Howell LL (2001) Compliant mechanisms. Wiley, New York

Howell LL, Magleby SP, Olsen BM (Hg.) (2013) Handbook of compliant mechanisms. Wiley, Chichester

Ivanov I, Corves B (2012) Flexure hinge-based parallel manipulators enabling high-precision micro manipulations. In: Ananthasuresh GK, Corves B, Petuya V (eds) Micromechanics and microactuators, vol 2, Mechanisms and machine science. Springer, Dordrecht, pp 49–60

Kong X, Gosselin C (2007) Type synthesis of parallel mechanisms, vol 33, Springer tracts in advanced robotics. Springer, Berlin/New York

Merlet J-P (2006) Parallel robots, vol 128, 2nd edn, Solid mechanics and its applications. Springer, Dordrecht

Raatz A (2006) Stoffschlüssige Gelenke aus pseudo-elastischen Formgedächtnislegierungen in Parallelrobotern (eng.: flexure hinges made of pseudo-elastic shape memory alloy in parallel robots), Schriftenreihe des Instituts für Werkzeugmaschinen und Fertigungstechnik der TU Braunschweig. Vulkan-Verl, Essen

Development of Adaptive Compliant Gripper Finger with Embedded Actuators

A. Milojević and N.D. Pavlović

Abstract Developing a gripper that can grasp objects of widely varying shapes represents a challenging task. Compliant mechanisms with embedded actuators are one way to obtain the gripper which could accommodate its grasping surface to any irregular and sensitive grasping object. This paper presents the new solution of adaptive compliant gripper finger with embedded actuators. Embedded actuators (within gripper finger structure) are able to morph shape of the grasping surface to accommodate different objects. Novel approach to the synthesis of compliant mechanisms with embedded actuators is also presented. The main contribution of this approach is that optimization of the actuator placement is included in the design process. It will be shown that gripper finger can achieve many complex grasping patterns via embedded actuators.

Keywords Adaptive gripper finger • Compliant mechanism • Embedded actuators • Shape morphing surface • Synthesis

1 Introduction

Developing a gripper that can grasp irregular and unpredictably shaped objects represents a challenging task. Grasping objects of widely varying shapes and surfaces essentially requires adaptability for safe and reliable gripping performance especially when fragile objects of different stiffness are manipulated. Much research has therefore been done to develop universal flexible grippers (Choi and Koc 2006; Jain et al. 2013; Jung et al. 2013; Giannaccini et al. 2014; Martinez et al. 2013; Ilievski et al. 2011; Araromi et al. 2014; Amend et al. 2012).

A flexible gripper design based on the use of compliant materials and internal pressure was introduced in (Choi and Koc 2006). This type of grippers conforms to the shape of an object by means of elastic gripping elements and pressurization. In (Jain et al. 2013) compliance is added in a selective compliant assembly robot arm (SCARA) in the form of a two ionic polymer metal composite (IPMC) fingers based

A. Milojević (✉) • N.D. Pavlović
Faculty of Mechanical Engineering, Department for Mechatronics and Control,
University of Niš, Niš, Serbia
e-mail: andrijamilojevic87@gmail.com; pavlovic@masfak.ni.ac.rs

E.-C. Lovasz et al. (eds.), *Microactuators and Micromechanisms*, Mechanisms and Machine Science 30, DOI 10.1007/978-3-319-15862-4_4

micro gripper. Paper (Jung et al. 2013) presented adaptive gripping device called the “Buckling gripper”. The key design principle of the Buckling gripper is inspired by a caterpillar’s proleg that highly deforms depending on the shape of the contact surface. This key principle is applied to the gripper via flexural buckling. In (Giannaccini et al. 2014) soft cable-driven gripper was introduced, featuring no stiff sections gripper is able to adapt to a wide range of objects due to its entirely soft structure. Paper (Martinez et al. 2013) presented soft tentacles based on micropneumatic networks spatially distributed at the interface of two different elastomers. These composite elastomeric structures enable complex 3D motion of the tentacles which was used to grip and manipulate objects with complex shapes. In (Ilievski et al. 2011) soft starfish gripper was introduced. The gripper was designed by using embedded pneumatic networks (Pneu-Nets) of channels in elastomers that inflate like balloons for actuation. By actuating the individual networks, the device could continuously change its shape from convex to concave and thus can grip different shaped objects. Paper (Araromi et al. 2014) presented the use of dielectric elastomer minimum energy structures (DEMES) as foldable and compliant gripper mechanisms. DEMES consist of a prestretched dielectric elastomer actuator membrane bonded to a flexible frame. The actuator finds equilibrium in bending when the prestretch is released and the bending angle can be changed by the application of a voltage bias. The proposed gripper design uses four multisegment DEMES actuator arms of length 30 cm in a cruciform configuration for secure target gripping and displays conformability to objects of varying shapes and sizes (Araromi et al. 2014). Further investigation led to a completely different approach to a universal gripper where individual fingers are replaced by a single mass of granular material that, when pressed onto a target object, flows around it and conforms to its shape (Amend et al. 2012). Most of these grippers require external drive (compressed air or electrical motor) (Choi and Koc 2006; Giannaccini et al. 2014; Martinez et al. 2013; Ilievski et al. 2011; Amend et al. 2012), high voltage (Araromi et al. 2014) and assembling. Moreover, there is no unique synthesis methodology.

One way to achieve adaptability is to use compliant mechanisms (Howell et al. 2013). A compliant mechanism can be defined as a single-piece flexible structure which uses elastic deformation to achieve force and motion transmission (compliant mechanisms deform smoothly as a whole) (Howell et al. 2013). This unique feature of compliant mechanisms is very promising option for the successful realization of shape-adaptable structures and provides a novel means to morph structural shape. Compliant mechanisms that change their shapes through structural deformations offer many benefits: reduced complexity, no wear, ease of manufacture, no assembly, better scalability, better accuracy, etc. Furthermore, compliant mechanisms can overcome many disadvantages of different mechanical grippers since the entire shape morphing structure is viewed as a compliant mechanism.

Synthesis of compliant mechanisms has been well studied in the past (Joo et al. 2000; Saxena and Ananthasuresh 2000), but little attention has been directed to problems related to synthesis of compliant mechanisms for shape morphing applications (Lu and Kota 2003, 2005; Zhao and Schmiedeler 2013). Synthesis methods developed in (Lu and Kota 2003, 2005; Zhao and Schmiedeler 2013) focuses on determination of optimal topology of compliant mechanisms so that they can achieve the target curve profile. But these papers only focus on problems regarding changing an initial shape to one target shape, where the final target shape is determined ahead of time and specified by the designer. These methods cannot be applied when developing a gripper that can adapt to unpredictably shaped objects. There are many situations where the desired shape change might not be known ahead of time and could be a function of the environment. For such applications, shape morphing gripper structures should be adaptive and should have a controllable response with exceptional manipulability.

This paper recasts the shape-changing problem as the challenge of how to design an adaptive compliant gripper that is controllable and capable of achieving multiple shapes. One such adaptive gripper could be formed as compliant mechanism with inherent actuators and sensors (Fig. 1). By embedding actuators and sensors within the compliant mechanism structure, gripper would be able to realize both sensing (via sensors) and appropriate response (via actuators and internal structure) to the unknown external environment, thus making the gripper adaptive. Such gripper is biologically inspired and holds the potential to lead to tightly integrated, highly functional, multi-purpose, adaptive system.

In this paper we pay more attention to the problem of embedding actuators since the primary goal of the paper is to develop gripper that is capable of achieving multiple grasping patterns i.e. to develop gripper that have structural adaptability. By embedding actuators within a compliant gripper structure, gripper may be capable of producing many complex deformations of the grasping surface i.e. gripper would be adaptive.

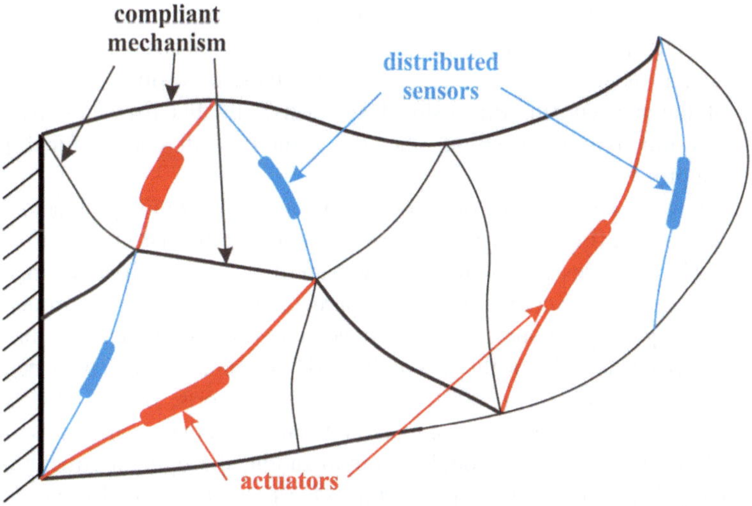

Fig. 1 Conceptual design of a compliant mechanism with embedded actuators and sensors

The algorithmic framework for distributed actuation and sensing within a compliant active structure has been already developed in (Trease and Kota 2009), but the proposed method often produces compliant mechanism with lot of intersections between elements as well as elements and actuators, which are very difficult (or nearly impossible) to manufacture. Beside this, intersection between elements often increases complexity and stiffness of the structure which can significantly lower the system functionality.

This paper presents the new solution of adaptive compliant gripper (in this paper only one gripper finger) with embedded actuators. We also present an improved design methodology for the simultaneous synthesis of compliant mechanism and actuator placement. The design methodology is improved so that compliant systems (compliant mechanism with embedded actuators) without intersecting elements are obtained (unlike solutions obtained in (Trease and Kota 2009)). This represents one novel approach to the synthesis of compliant systems. We will demonstrate that obtained adaptive compliant gripper with embedded actuators may be capable of achieving many different grasping patterns and thus having many advantages over existing grippers.

2 Synthesis Methodology

To develop adaptive compliant gripper we use compliant mechanisms with distributed compliance (Joo et al. 2000), since the distributed compliance throughout the compliant mechanism provides a smooth deformation field, which reduces the stress concentration. The continuum synthesis approach is usually used for the design of mechanisms with distributed compliance (Joo et al. 2000; Saxena and Ananthasuresh 2000; Milojević et al. 2013). The synthesis methodology used in this approach involves two stages: generation of the mechanism topology and determination of optimum size, geometry, and shape of various constituent elements of the mechanism (dimensional synthesis). In this paper we pay more attention to topology optimization, since this is more creative and difficult part of the design process.

The allowable space for the design in a topology optimization problem is called the design domain (Fig. 2a). The topology is defined by the distribution of material and void within the design domain or as the pattern of connectivity of elements in a structure (Fig. 2b). The continuum based approach focuses on the determination of the optimal topology (the best material connectivity in a compliant structure). The designer only needs to define the size of the design domain in which the mechanism should fit, location of the supports, input and output ports, size of applied loads (Fig. 2a) as well as properties of the material from which the mechanism should be produced. Then, through the topology optimization, the optimal structural form (optimal topology) of a compliant mechanism for a specified input force and output deflection requirements is automatically generated (Fig. 2b).

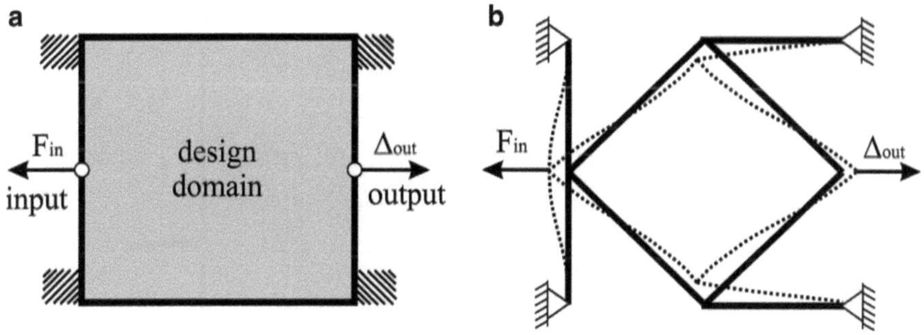

Fig. 2 Synthesis of compliant displacement inverter in which the input force F_{in} and the output displacement Δ_{out} are in opposite directions: (**a**) design domain; (**b**) optimal topology of compliant displacement inverter (deformed position is shown with *dash lines*) (Lu and Kota 2005)

Adaptive compliant gripper could be seen as compliant mechanism with embedded actuators; embedded actuators would provide the gripper with ability to achieve multiple grasping patterns, thus making the gripper adaptive. In classical synthesis approach, actuators are added after a compliant mechanism is already developed, where determination of actuator placement is occurring outside the optimization process at the designer's option. In the literature the synthesis of single-actuator mechanisms has mainly been considered (Joo et al. 2000; Saxena and Ananthasuresh 2000; Lu and Kota 2003, 2005; Zhao and Schmiedeler 2013).

To develop an adaptive gripper, the structural topology of a compliant mechanism and actuator placement must be simultaneously synthesized (actuator type, orientation, size and location must be integrated as variables in the design process). Here actuator placement affects the structural topology of a mechanism and vice versa. Moreover, multiple actuators must be used. When multiple actuators are used, the question of control of such system must be considered. Incorporating control during the optimization process can enhance the controllability of an adaptive system. Central to this method is the concept of structural orthogonality, which refers to the unique system response for each actuator it contains (Trease and Kota 2009). These concepts (controllability and structural orthogonality) represent the structural adaptability of the system and are explored in detail in (Trease and Kota 2009).

In (Milojević et al. 2013) we have developed computer-coded algorithm for synthesis of compliant mechanisms with distributed compliance, and improved the topology optimization technique in (Milojević and Pavlović 2013). To develop adaptive compliant gripper we modify our computer-coded algorithm so that actuator placement is also included in the synthesis process. Actuators, modeled as both force generators and structural compliant elements, are included as topology variables in the optimization. We also incorporate control in the synthesis process through the use of structural orthogonality concept.

The outline of the methodology that we use for developing adaptive compliant gripper with embedded actuators is shown in a pictorial example (Fig. 3). First the

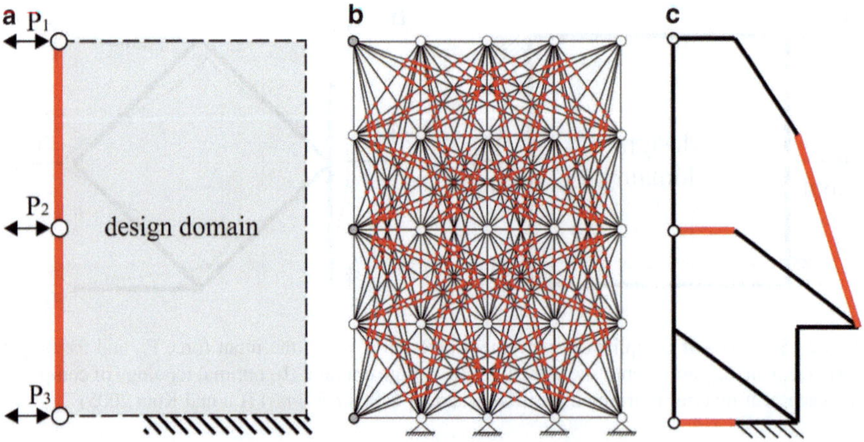

Fig. 3 The steps in the synthesis methodology: (**a**) problem specifications; (**b**) parameterization (intersections between elements are indicated by *red dots*); (**c**) optimized topology of adaptive compliant gripper finger with embedded actuators (actuators are indicated by *red lines*)

Table 1 Design specification for developing adaptive compliant gripper with embedded actuators

Design parameters	
Design domain	120 mm × 80 mm
Grid size	5 × 5
Degree of nodal connectivity	4
Number of beam elements	168
Element modulus	$E_{el} = 2.48$ GPa
Actuator modulus	$E_{akt} = 500$ MPa
Actuator block force	90 N
External load	0.1 N
Element out-of-plane thickness	1.5 mm
Element thickness choice	0.5, 1, 1.5 mm
Thickness of shape morphing surface	0.5 mm

problem specifications are defined only for one gripper finger (Fig. 3a and Table 1). These includes: size of the design domain, grasping surface (left boundary of the design domain) and number of output points (three output points are chosen to represent the output region where horizontal direction is set to be the desired direction of the output deflection), supports (the bottom boundary of the design domain – one part only), property of the material (Young modulus) from which the mechanism should be built and other constraints such as minimum value of the output deflection $d^{act}_{\text{min,target}}$ and total element length L_t which is equivalent to the volume constraint. All the design parameters are given in Table 1.

Next the design domain is parameterized (Fig. 3b). The physical design space must be broken down so as to be represented by a set of variables that an optimizer can act on. Since we use compliant mechanisms with distributed compliance (Milojević and Pavlović 2013), the Grounded Structure Approach (GSA) (Joo et al. 2000; Saxena and Ananthasuresh 2000; Milojević et al. 2013) is used for the parameterization. Therefore, the prescribed design domain is divided into a number of nodes, and a network of beam elements connecting these nodes serves as an initial guess. The design variables are the thickness of each element and variable that marks the element selected to be actuator in a given structure; this variable has a value between 1 and the total number of elements (Table 1). The linear actuator model is used where at the ends of the beam axial force, equal to the block force (Trease and Kota 2009), is applied. A thickness value of zero deactivates the element, removing it from the structure; other values represent thickness values (Table 1). It is important to note that we used partially connected ground structure (Fig. 3b) i.e. the ground structure that is not "fully connected" (not all the nodes in the ground structure are interconnected). A fully connected structure can lead to overlapping elements that are difficult to produce. Thus, certain filters (as - computer-code) are used to eliminate the overlapping elements. In addition, the degree of nodal connectivity (Milojević and Pavlović 2013) is defined as well (Table 1).

When using the partially connected ground structure the solutions with intersecting elements are obtained (Joo et al. 2000; Milojević et al. 2013) where elements and actuators would intersect also (Trease and Kota 2009). Producing a structure with intersecting elements as well as actuators is very difficult. Moreover intersections between elements often increase complexity and stiffness of the structure which can significantly lower the system functionality. This deficiency has motivated us to improve the existing topology optimization technique (Trease and Kota 2009) so that the intersections between elements as well as actuators would be eliminated in the process of optimization. We apply the same ide as in (Milojević and Pavlović 2013), but now for the problem of synthesis of adaptive compliant gripper with embedded actuators. All the parameters regarding the parameterization are given in Table 1.

After the parameterization is done, search method is applied to find the optimal compliant mechanism with embedded actuators. Because of the broad design space and number of elements, topology synthesis problems are solved with optimization methods. The goal of the optimization in the synthesis of compliant mechanisms with embedded actuators is to minimize the actuator number and maximize structural adaptability of a compliant system (maximize controllability (Trease and Kota 2009)) while meeting given constraints. By maximizing controllability gripper will be able to achieve multiple grasping patterns of its shape morphing surface. Three actuators are required minimally to fully control the three output points (Fig. 3a). Thus, the number of actuators is held constant during the optimization, rather than minimized. To achieve a desired goal in the process of optimization, an objective function is needed. The objective function we used for the synthesis of the adaptive compliant gripper with embedded actuators is:

$$\textbf{maximize}\begin{bmatrix} \eta_C - w_1 \cdot \left(d^{act}_{min,target} - d^{act}_{min}\right) - w_2 \cdot \left|d^{ext}_{max}\right| - \\ -w_3 \cdot \left|L_t - L_{target}\right| - w_4 \cdot n_{int} \end{bmatrix}. \quad (1)$$

where: η_C is controllability (Trease and Kota 2009), w_1, w_2, w_3 and w_4 are relative weighting constants, $d^{act}_{min,target} = 1$ mm is the minimum value of the output deflection, d^{act}_{min} is the displacement of the output points (Trease and Kota 2009), d^{ext}_{max} is the displacement of the output points caused by external loading (the external load is applied at all three output points, all at once) (Trease and Kota 2009), L_t is total element length, $L_{target} = 721.1$ mm is the desired total element length (Trease and Kota 2009) and n_{int} represent the total number of intersections in the structure (Milojević and Pavlović 2013).

All objective function terms and constraints are calculated from the results of the linear finite element analysis (FEA), implemented in the computer-coded algorithm.

When the parameterization is discrete i.e. elements are either on or off, the discrete optimizations methods are used, of which Genetic Algorithms (GA) (Reeves and Rowe 2002) are applied here. The genetic algorithm parameters used in the synthesis of the adaptive compliant gripper are: initial population of 200 designs, total number of 1,000 generations, roulette selection function, cross-over probability of 95 %, elite count of 2 members, and mutation probability of 9 %.

To obtain the adaptive compliant gripper with embedded actuators more than 20 GA's are run. The optimization process starts with 168 beam variables and total number of 1,664 intersections in the initial ground structure (Fig. 3b). Figure 3c shows the result for the best of population of 210 over 1,000 generations. This obtained solution contains the compliant structure in which some of the elements are eliminated and some chosen as actuators in the process of optimization. The remaining elements together with elements selected to be actuators define the optimal topology of the adaptive compliant gripper finger. High controllability ($\eta_C = 97.57$ %) of a compliant system is achieved. The results demonstrate that all the constraints are satisfied as well. Unlike solutions in (Trease and Kota 2009) here the compliant system without intersections is obtained (Fig. 3c). This means that totally 1,664 intersections are eliminated in the process of optimization.

3 FEM Analysis of the Adaptive Compliant Gripper Finger with Embedded Actuators

Based on the obtained solution (Fig. 3c) 3D solid model of the adaptive compliant gripper finger with embedded actuators was designed (Fig. 4a). Instead of using real actuators, here the actuators were modeled as thin elastic elements (in a form of a

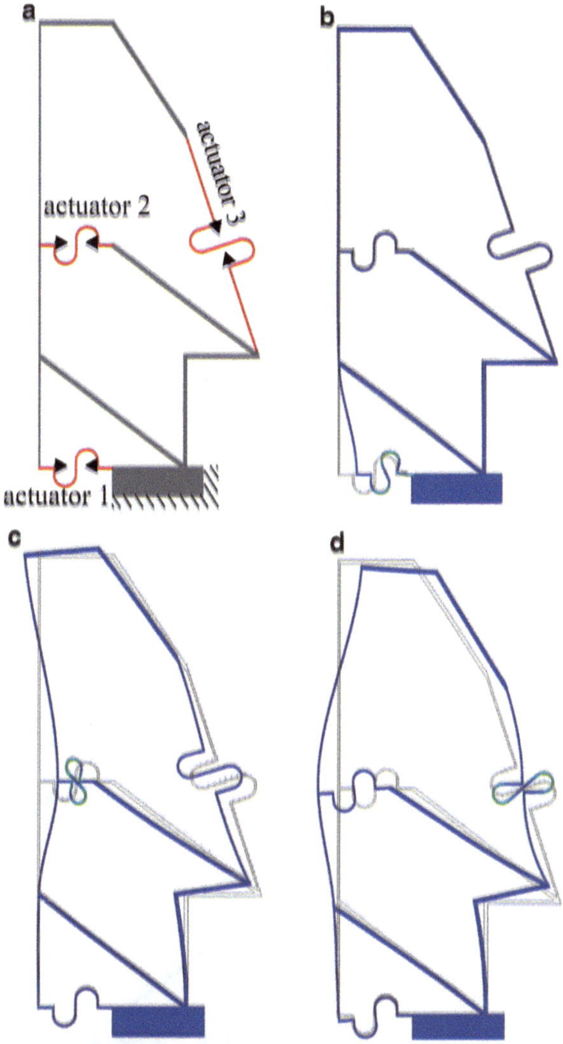

Fig. 4 Design of adaptive compliant gripper finger with embedded contracting actuators (**a**) and FEM results when actuator 1 (**b**), actuator 2 (**c**) and actuator 3 (**d**) is active

spring) that allow the actuation (Fig. 4a). We decided to model only one finger of the gripper as the fingers in two-fingered or multi-fingered gripper would have the same behavior.

To investigate adaptability of the compliant gripper finger to different shapes of the gripping objects FEM simulations were performed (ABAQUS software was used). As boundary condition fixed support was applied at immobile part of the gripper finger (Fig. 4a). To simulate the contraction of the actuators, as input a displacement of ± 5 mm (stroke of actuator) in the direction of the actuators axis was introduced at the both ends of the all actuators (Fig. 4a) while the vertical displacement is set to be 0. In this paper both actuators that contract and actuators that extend are simulated. To show the capability of the gripper finger to produce

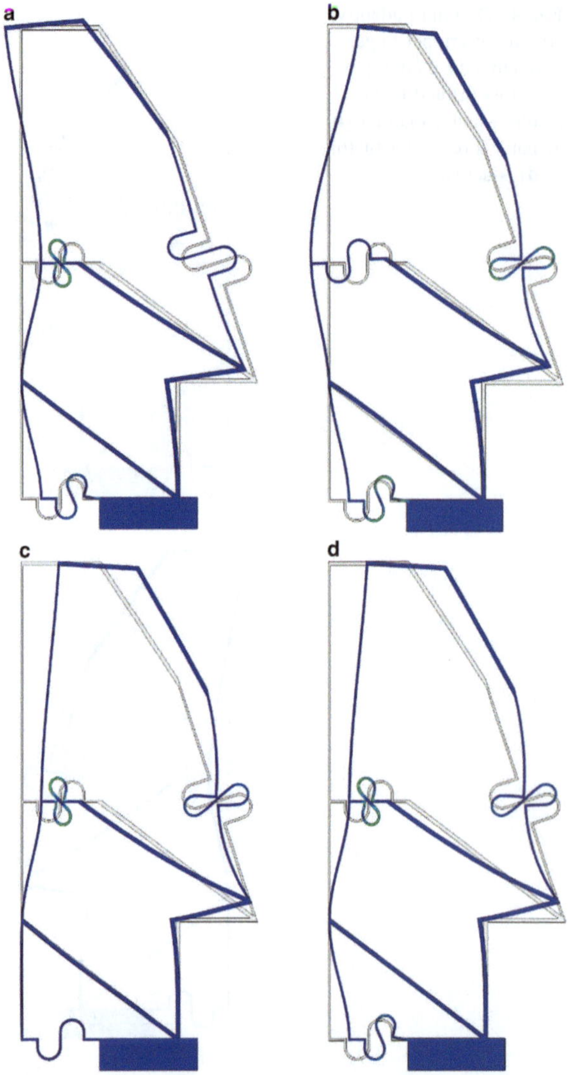

Fig. 5 FEM results for combination of different active actuators: (**a**) actuators 1 and 2; (**b**) actuators 1 and 3; (**c**) actuators 2 and 3; (**d**) actuators 1, 2 and 3

multiple shapes of its shape morphing surface the FEM simulations were performed without any grasping object. Figure 4 shows the results when contracting actuators are used.

The FEM results show that two main grasping patterns could be created: 'convex' (Fig. 4c) and 'concave' patterns (Fig. 4d); when actuator 2 is active an object of convex shape could be grasped (Fig. 4c), and when actuator 3 is active an object of concave shape could be grasped (Fig. 4d). Also actuator 1 could be used to accommodate the grasping surface of the gripper finger to different irregular shapes of the grasping object. Different grasping patterns could be achieved with different combinations of active actuators (Fig. 5). The results (Fig. 5) show that gripper

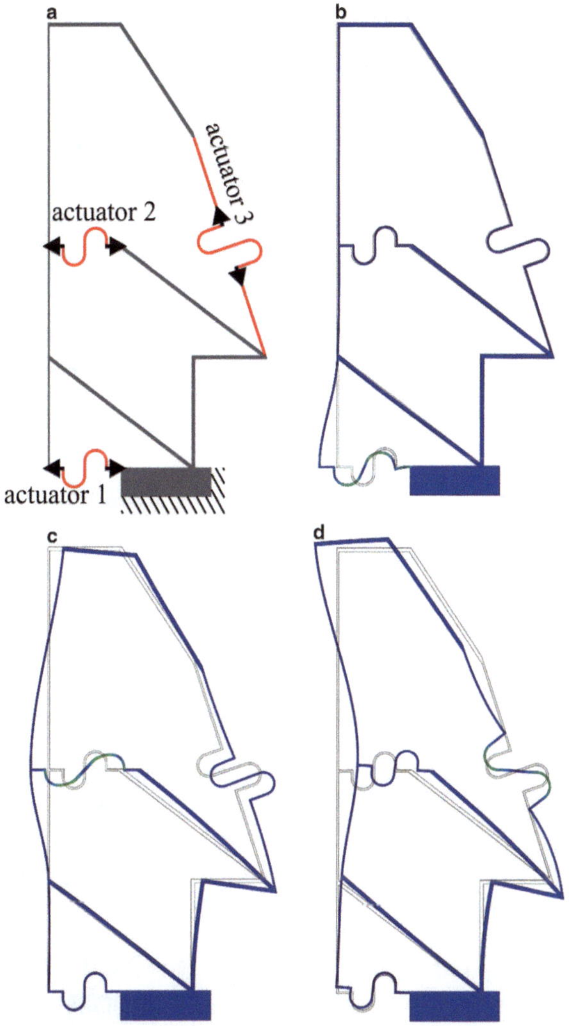

Fig. 6 Design of adaptive compliant gripper finger with embedded extending actuators (**a**) and FEM results when actuator 1 (**b**), actuator 2 (**c**) and actuator 3 (**d**) is active (Milojević et al. 2014)

finger has capability to grasp objects of different shapes and sizes, for example when using combination of actuator 1 and actuator 2 concave-convex objects could be grasped (Fig. 5a).

Another set of FEM simulations were performed, but now with extending actuators (Milojević et al. 2014). The FEM results are shown in Fig. 6. The same as in the previous case (when contracting actuators are used) two main grasping patterns could be created: 'concave' patterns (Fig. 6c) and 'convex' (Fig. 6d). The difference is that now when actuator 3 is active an object of convex shape could be grasped (Fig. 6d), and when actuator 2 is active an object of concave shape could be grasped (Fig. 6c). The FEM results for combination of different active contracting actuators are shown in Fig. 7.

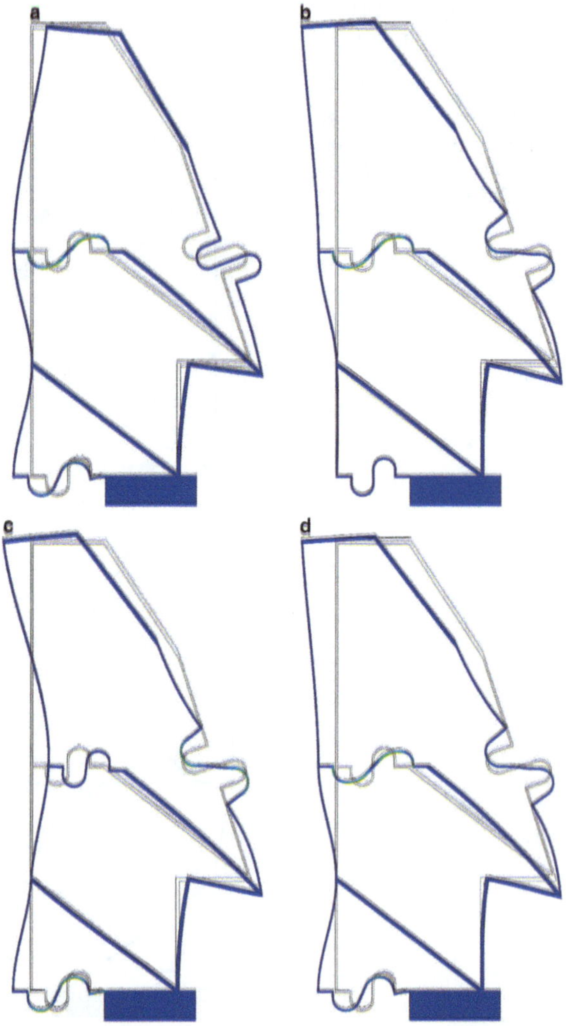

Fig. 7 FEM results for combination of different active extending actuators: (**a**) actuators 1 and 2; (**b**) actuators 2 and 3; (**c**) actuators 1 and 3; (**d**) actuators 1, 2 and 3 (Milojević et al. 2014)

Combinations of extending and contracting actuators could be used to achieve a variety of different shapes of the gripper finger shape morphing surface. Some characteristic shapes that could be grasped with combinations of actuators that extend and actuators that contract are shown in Fig. 8.

For different stroke of actuators, objects of different size could be grasped. For example, when actuator 2 is active (Fig. 4c) for different stroke of actuator (1, 2 mm, ...) convex objects of different radius could be grasped (Fig. 9). This is also the case when actuators 1 and 3 are active and with different combinations of active actuators.

In the design of the gripper finger presented in this paper, actuators with stroke larger than 5 mm (at both ends) could be used; the stroke is limited by the space in

Fig. 8 FEM results for combination of different active extending and contracting actuators: (**a**) actuators 1 (extending) and 2 (contracting); (**b**) actuators 1 (extending) and 3 (contracting); (**c**) actuators 1 (extending), 2 (contracting) and 3 (contracting); (**d**) actuators 1 (contracting), 2 (extending) and 3 (extending)

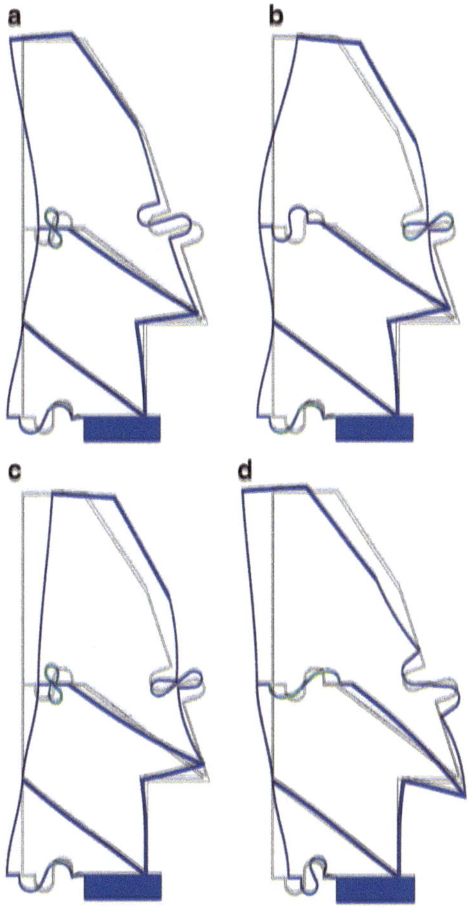

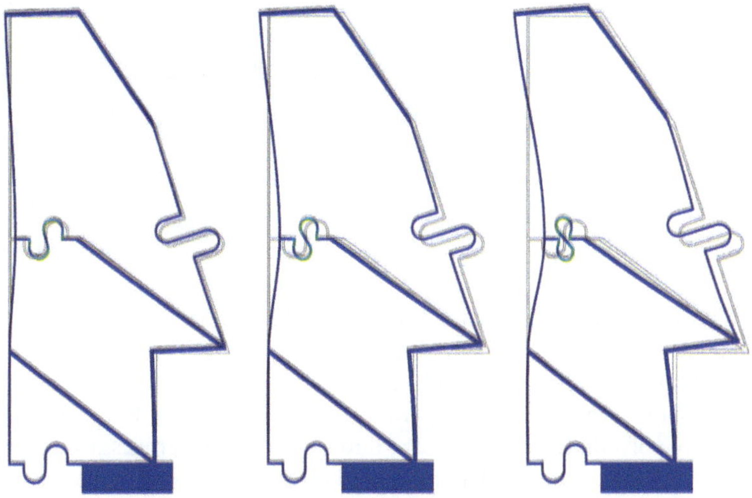

Fig. 9 Objects of different radius that could be grasped when actuator 2 is active

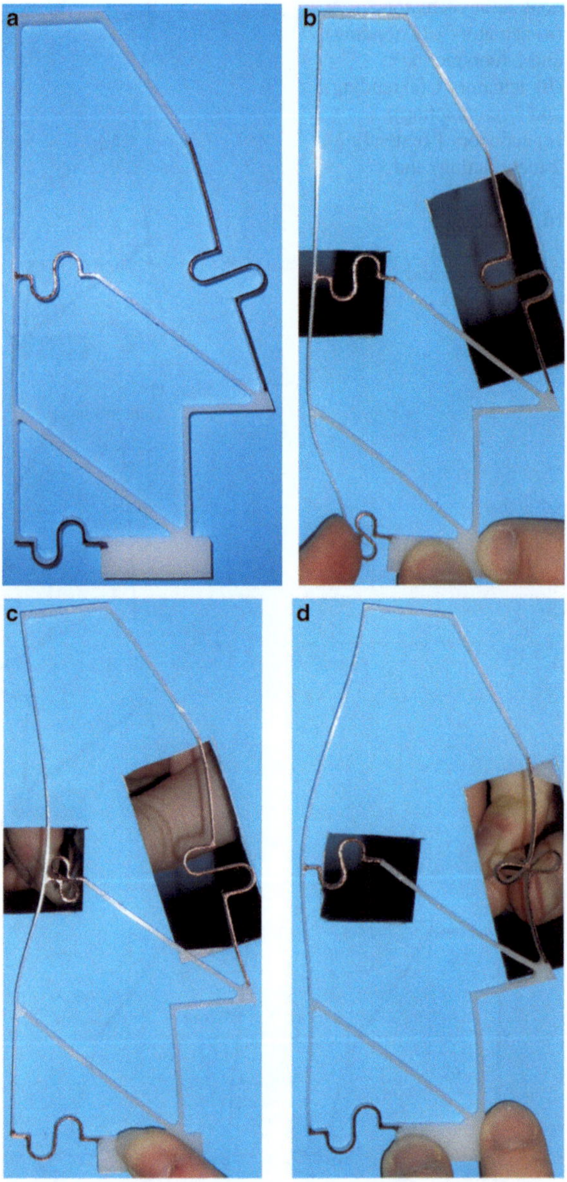

Fig. 10 Produced gripper finger (**a**) and its adaptable behavior when contracting actuator 1 (**b**), actuator 2 (**c**) and actuator 3 (**d**) is active

which actuator should fit. Actuator 1 and 2 could have the stroke of up to 10 mm (at both ends) and actuator 3, up to 60 mm.

To verify the FEM results, physical prototype of the adaptive compliant gripper finger was made (Fig. 10). The gripper finger model was made from HDPE plastic (high density polyethylene E = 0.5 GPa) throughout milling process using the CNC machine. The same as in simulations instead of embedding real actuators we used

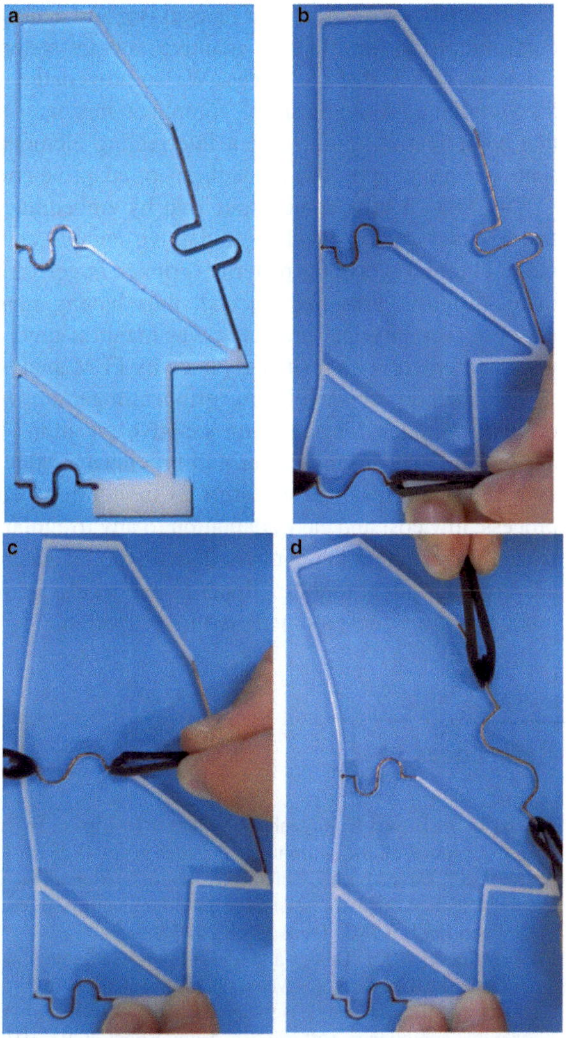

Fig. 11 Produced gripper finger (**a**) and its adaptable behavior when extending actuator 1 (**b**), actuator 2 (**c**) and actuator 3 (**d**) is active (Milojević et al. 2014)

thin elastic elements (in form of a spring). Figure 10 shows the produced gripper finger. The adaptable behavior of produced gripper finger (Figs. 10 and 11) is in good agreement with the results obtained by FEM analysis.

4 Conclusions

This paper presents the adaptive compliant gripper finger which has capability to adapt to different grasping objects via embedded actuators. With the introduction of multiple, optimally placed actuators our work focuses on developing gripper that has ability to achieve any shape required of the gripper.

The synthesis of compliant mechanism and actuators in one system is the main contribution in developing adaptive robotic gripper finger that can achieve unknown target shape changes. We improved the existing design methodology for simultaneous synthesis of compliant mechanism and actuator placement so that compliant systems without intersecting elements were obtained. This represents one novel approach to synthesis of adaptive compliant systems.

The FEM results demonstrate that by embedding actuators within the gripper finger structure the developed adaptive compliant finger has ability to produce many complex grasping patterns (gripper finger can accommodate its surface to many different grasping objects), thus having many advantages over existing grippers. The physical prototype was manufactured and tested. The results are in good agreement with results obtained by FEM analysis.

The method for developing gripper finger utilized in this paper is new and original principle for achieving adaptive grasping. The main difference between design presented in this paper and previously established gripper lies in: gripper structure and embedded actuators as monolithic structure, simple manufacturing process, low cost and easy adaptation to any irregular object shape. It is possible to make this gripper finger in micro domain too. In general, the results indicate that the adaptive compliant robotic gripper finger developed in this paper is a good starting point for further development and investigation.

References

Amend JR Jr, Brown E, Rodenberg N, Jaeger H, Lipson H (2012) A positive pressure universal gripper based on the jamming of granular material. IEEE Trans Robot 28:341–350

Araromi OA, Gavrilovich I, Shintake J, Rosset S, Richard M, Gass V, Shea HR (2014) Roll-able multisegment dielectric elastomer minimum energy structures for a deployable microsatellite gripper. IEEE/ASME Trans Mech 20:438–446

Choi H, Koc M (2006) Design and feasibility tests of a flexible gripper based on inflatable rubber pockets. Int J Mach Tools Manu 46:1350–1361

Giannaccini ME, Georgilas I, Horsfield I, Peiris BHPM, Lenz A, Pipe AG, Dogramadzi S (2014) A variable compliance, soft gripper. Auton Robot 36:93–107

Howell LL, Magleby SP, Olsen BM (2013) Handbook of compliant mechanisms. Wiley, Somerset

Ilievski F, Mazzeo AD, Shepherd RF, Chen X, Whitesides GM (2011) Soft robotics for chemists. Angew Chem Int Ed 50:1890–1895

Jain RK, Majumder S, Dutta A (2013) SCARA based peg-in-hole assembly using compliant IPMC micro gripper. Robot Auton Syst 61:297–311

Joo J, Kota S, Kikuchi N (2000) Topological synthesis of compliant mechanisms using linear beam elements. Mech Struct Mach 28(4):245–280

Jung G-P, Koh J-S, Cho K-J (2013) Underactuated adaptive gripper using flexural buckling. IEEE Trans Robot 29(6):1396–1407

Lu K-J, Kota S (2003) Design of compliant mechanisms for morphing structural shapes. J Int Mater Syst Struct 14:379–391

Lu K-J, Kota S (2005) An effective method of synthesizing compliant adaptive structures using load path representation. J Int Mater Syst Struct 16:307–317

Martinez RV, Branch JL, Fish CR, Jin L, Shepherd RF, Nunes RMD, Suo Z, Whitesides GM (2013) Robotic tentacles with three-dimensional mobility based on flexible elastomers. Adv Mater 25:205–212

Milojević A, Pavlović ND (2013) Software for synthesis of compliant mechanisms without intersecting elements. Facta Universitatis Series Mech Eng 11(2):153–168

Milojević A, Pavlović ND, Milošević M, Tomić M (2013) New software for synthesis of compliant mechanisms. In: Proceedings of the 2nd international conference mechanical engineering in XXI century, Niš, pp 273–278

Milojević A, Pavlović ND, Pavlović NT (2014) Adaptive compliant gripper finger with embedded extending actuators. In: Proceedings of the 58th Ilmenau Scientific Colloquium IWK, vol 2119, Ilmenau, p 11

Reeves CR, Rowe JE (2002) Genetic algorithms-principles and perspectives: a guide to GA theory. Kluwer Academic Publishers, Dordrecht

Saxena A, Ananthasuresh GK (2000) On an optimal property of compliant topologies. Struct Multidisc Optim 19:36–49

Trease B, Kota S (2009) Design of adaptive and controllable compliant systems with embedded actuators and sensors. J Mech Design 131(111001):12

Zhao K, Schmiedeler JP (2013) Using rigid-body mechanism topologies to design shapechanging compliant mechanisms. Proceedings of the 2013 ASME international design engineering technical conferences and computers & information in engineering conference, Portland, 4–7 Aug. http://ame-robotics.nd.edu/LAB/Publications/DETC2013-12576-preprint.pdf

Overview and Classification of Flexure Hinge Based Micromanipulators

D. Schoenen, I. Prause, S. Palacios, and B. Corves

Abstract It is difficult to find a comprehensive classification containing all aspects of robots, not to mention of micromanipulators, in present literature. With the focus on design and kinematic attributes, the objective of this paper is to create a classification of robots which can also be applied for micromanipulators. The different attributes are identified and discussed. Existing micromanipulators are grouped according to this classification scheme and a clear overview of their specific structural characteristics and performance is given. Thus three different groups of micromanipulators can be identified.

Keywords Micromanipulator • Overview • Classification • Compliance • Flexure hinge

1 Introduction

General technological development and better knowledge of compliant mechanisms increase the growth in compliant mechanisms applications (Howell et al. 2013). This is due to many advantages of compliant mechanisms like the monolithic structure, which induces reduced assembly, reduced wear, reduced need for lubrication and weight reduction. This enables not only close to zero backlash and frictionless motion but also significant potential for miniaturization and increased performance such as, e.g. high precision (Howell 2001). The disadvantages are undesired rotational and/or translational parasitic effects, deviation in motion, difficulties in analyzing and designing compliant mechanisms and the limited range of motion (Howell 2001). But for high precision parallel micromanipulators with a workspace typically in the millimeter range and macroscopic dimensions the mentioned disadvantages can be compensated by the use of precision flexure hinges. Thus the main focus of this paper lies on flexure hinge based micromanipulators.

D. Schoenen (✉) • I. Prause • S. Palacios • B. Corves
Institut für Getriebetechnik und Maschinendynamik, RWTH Aachen University, Aachen, Germany
e-mail: Schoenen@igm.rwth-aachen.de; prause@igm.rwth-aachen.de; corves@igm.rwth-aachen.de

E.-C. Lovasz et al. (eds.), *Microactuators and Micromechanisms*, Mechanisms and Machine Science 30, DOI 10.1007/978-3-319-15862-4_5

In Sect. 2 a general classification of robots, especially for micromanipulators, is introduced. The classification scheme focuses mainly on attributes of the design and kinematics of robots. The classification is applied for existing micromanipulators and specific characteristics of these micromanipulators are listed in Sect. 3. Finally, Sect. 4 gives a brief conclusion of the paper and a perspective on further research projects.

2 Classification According to Design and Kinematics

In this chapter a classification of robots, especially of micromanipulators, is introduced. In general, robots can be classified according to their design and kinematics without considering their application and function. Important aspects are shown in Fig. 1 and are discussed in more detail in the following paragraphs. The classification scheme is based on (Bleicher 2003; Cao et al. 2013; Corves and Müller 2013; Haun 2007; Naval 1989; Röse 2011) and on analyses of existing mechanisms. To show the differences between a classical robot and a micromanipulator, those parts of the classification which are relevant for micromanipulators are highlighted.

2.1 Components of a Robot

As shown in Fig. 1 there are various important components of a robot. The *mechanical structure* includes the frame, limbs, joints and other necessary mechanical parts like springs or dampers. Limbs are connected to each other and to the frame through joints. Joints give the structure a certain degree of freedom and allow translational and/or rotational motion between the limbs. Basic joints are the *prismatic joint* (P) and the *revolute joint* (R). Further joints are combinations of these joints, for example, the *universal joint* (U) or the *spherical joint* (S). One of the benefits of a flexure hinge based micromanipulator as aforementioned is the monolithic structure of the limbs. Thus the different components of the mechanical structure of a micromanipulator are linked permanently without any backlash and cannot be disassembled like structures with conventional joints. Nevertheless, joints and limbs can be distinguished as long as the compliances of the structure are concentrated at certain points, representing the joint axes. The *actuator* transforms the provided power into a translational or rotational motion. For this reason, actuators are basically divided into rotary and linear actuators. Furthermore, actuators can be classified according to the form of energy that is used to generate a motion. Because of the limited range of motion of flexure hinges it is not possible to actuate a rotational axis directly in an appropriate accuracy range. Thus rotational axes of high precision micromanipulators are mostly driven by precise translational actuators in combination with a lever arm. The *end-effector* is the last element of the kinematic chain of a robot and is supposed to interact with its environment.

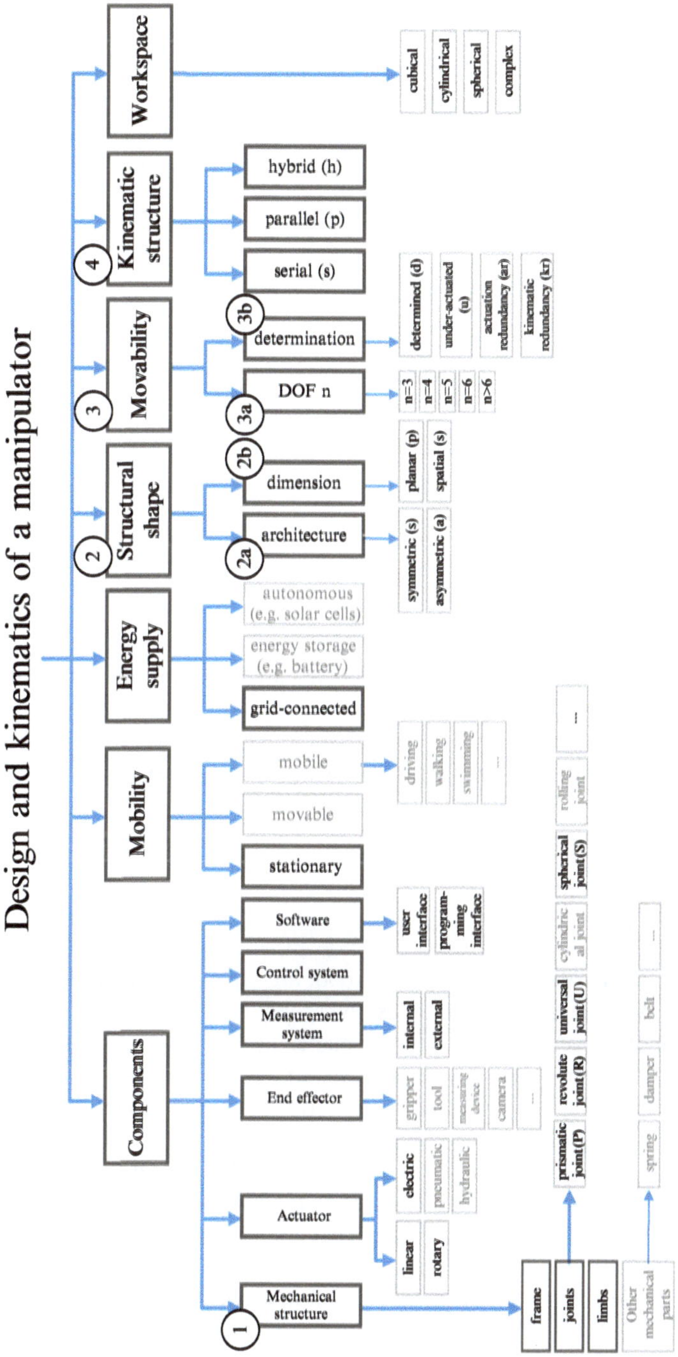

Fig. 1 Design and kinematics of a manipulator, applied for micromanipulators

As micromanipulators are used for high precision positioning, generally no special devices such as grippers or other tools are necessary. The *measurement system, the control system* and the *software* are important parts of a functioning robot, but not that important for the classification, because these aspects can be changed independently from the design and kinematics of a robot. Depending on the intended application, occasionally the electronic components of a micromanipulator have to be suitable for vacuum or other harsh environments.

2.2 *Mobility*

According to (Naval 1989), robots can be classified into *stationary*, *movable* and *mobile* robots. High accurate micromanipulators are delicate and sensitive due to the small aspect ratios of the flexure hinges. Therefore, such systems are not used for movable applications.

2.3 *Energy Supply*

For classical robots the energy supply can take various forms. Stationary robots are conventionally *grid-connected*. Movable or mobile robots need an *energy storage* or *autonomous* internal energy supply. Micromanipulators are primarily used stationary and therefore, a grid-connected energy supply is suitable.

2.4 *Structural Shape*

Two aspects of the structural shape of a robot are regarded: the *dimension* and the *architecture*. With regard to the dimension, a robot can be *planar* (p) or *spatial* (s). A planar robot moves in a pre-defined plane. In this case, all motion axes are parallel to each other. A spatial robot moves in a three-dimensional space, not all motion axes are parallel. A special form of a spatial robot is a spherical robot whose motion axes intersect all in one common point. Furthermore, a robot has either a *symmetric* (s) or an *asymmetric* (a) architecture. A symmetrical mechanism is composed of a moving platform connected to the base with geometrical symmetry by several identical legs. Mechanisms that do not meet the above-mentioned criteria are asymmetric. This classification can be applied to parallel mechanisms (Cao et al. 2013) and is valid also for parallel micromanipulators.

2.5 *Movability*

The movability of a robot, in particular also of a micromanipulator, is primarily defined by the *degree of freedom* (DoF) and the *determination* of the mechanism. If the degree of freedom is equal to the number of actuators, the robot is *determined* (d). Sometimes it is useful to design a *redundant* (r) or an *underactuated* (u) mechanism, i.e. the degree of freedom is different to the number of drives. An underactuated robot has less drives than necessary for a determined mechanism, whereas a redundant robot exhibits more actuators than its degree of freedom (*actuation redundancy* (ar)) (Garg et al. 2009; Pierrot 2002). The other type is the *kinematic redundancy* (kr). Further kinematic chains, possessing the same degree of freedom as the entire structure, are assembled additionally to increase the stiffness.

2.6 *Kinematic Structure*

There are two possible kinematic structures of robots: *serial* (s) and *parallel* (p). The combination of these two structures results in a *hybrid* (h) structure. Serial kinematic structures consist of open kinematic chains, whereas parallel structures exhibit closed kinematic chains. Only parallel and hybrid structures can be found among micromanipulators. This is due to the already mentioned fact that rotational axes cannot be actuated directly.

2.7 *Workspace*

Each robot has a particular workspace. The workspace defines the space the robot is able to reach in order to perform all desired tasks. The size and the shape of the workspace depend on the design and kinematic structure of the robot.

3 Overview of Developed Micromanipulators

The existing micromanipulators are grouped according to the classification made in the previous chapter and the main characteristics are specified in Table 1 (Please see Fig. 1 for the used abbreviations).

Firstly, the micromanipulators are grouped according to the place of development. There are four important universities who already manufactured prototypes of micromanipulators: the École Polytechnique Fédérale de Lausanne in Switzerland (I), Nanyang Technological University in Singapore (II), National Taiwan University in Taipei (III) and the Technische Universität Braunschweig in Germany (IV).

Table 1 Overview of developed micromanipulators

Place of development (University)	Name of the micro-manipulator	Mechanical structure 1	Architecture 2a	Dimension 2b	Translational DoF (T) 3a	Rotational DoF (R) 3a	determination 3b	Kinematic structure 4	Reference
I	Agietron micro-nano	3-$\underline{P}$RPa$^{4U*)}$	s	s	3	0	d	p	Lubrano and Clavel (2010)
	Orion MinAngle	3-$\underline{P}$SR	s	s	1	2	d	p	Pham et al. (2005)
	Sigma 6	6-$\underline{P}$US	s	s	3	3	d	p	Fazenda et al. (2006)
	Delta^3	3-$\underline{P}$Pa$^{4U*)}$	s	s	3	0	d	p	Beltrami et al. (2004)
II	XYZ FPM 3-PPP	3-$\underline{P}$PP	a	s	3	0	d	p	Tang et al. (2006)
	XYZ FPM 3-PRRR	3-$\underline{P}$RRR	s	s	3	0	d	p	Pham and Chen (2006)
	3PU FPM	3-$\underline{P}$U	s	s	1	2	d	p	Yang et al. (2011)
	3RRR micromotion stage	3-$\underline{R}$RR	s	p	2	1	d	p	Lu et al. (2004)
III	XYΘ_Z micropositioner	3-$\underline{P}$RRR	s	p	2	1	ar	h	Shuo Hung Chang et al. (1999)
	6-PSS nanopositioner	6-$\underline{P}$SS	s	s	3	3	d	p	Tung-Li Wu et al. (2008)
IV	MICABO^{HS}	–	s	s	3	3	d	h	Raatz (2006)
	MICABO^{ES}	3-$\underline{P}$RR	s	p	2	1	d	p	Raatz (2006)
	TRIGLIDE^S	3-$\underline{P}$RPa^{4R}R$^{*)}$	s	s	3	0	d	p	Raatz (2006)

$^{*)}$Pa represents a parallel substructure

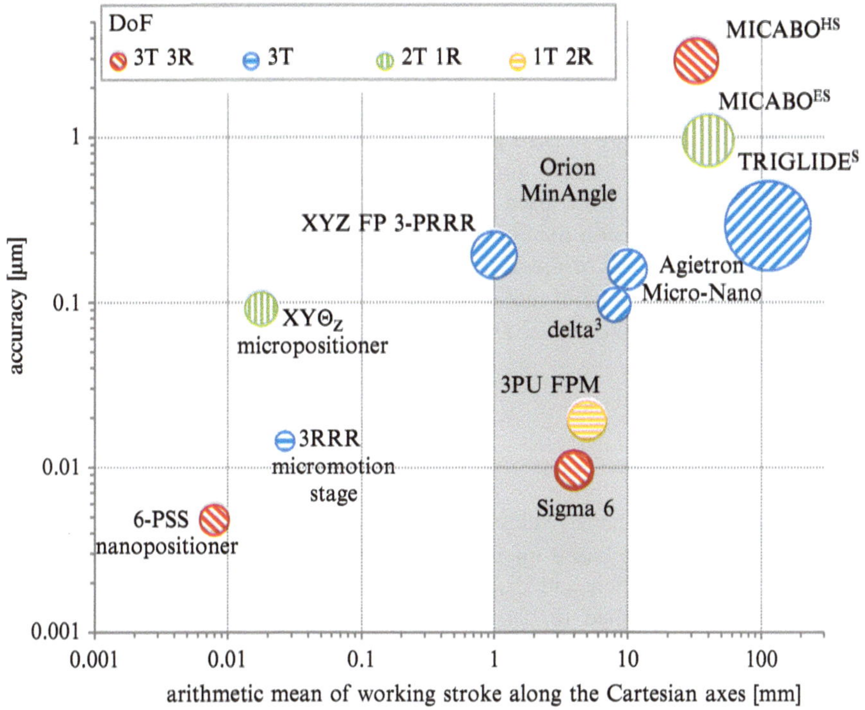

Fig. 2 Performance comparison of developed micromanipulators (the arithmetic mean of the edge length of the installation space is represented by the area of the *circle*)

The list also provides – as far as possible – information concerning the mechanical structure (1). An underlined letter corresponds to the actuated joint within one limb. The structural shape of the manipulators is described by the architecture (2a) and dimension (2b). Both the translational and rotational DoFs (3a) are specified to describe the movability together with the degree of determination (3b). In addition, the kinematic structure is specified (4).

To compare the performance of different manipulators the accuracy is plotted against the arithmetic mean of the working stroke along the Cartesian axes of the manipulator's workspace with double logarithmical scale (cf. Fig. 2). This allows the comparison of planar and spatial structures. The arithmetic mean of the edge length of the installation space of the micromanipulators is the third dimension of the plot and is represented by the area of the circle. The information about the accuracy of the different manipulators varies. Figure 2 is only a qualitative comparison, since some authors provide values for the absolute accuracy and others use the repeatability to describe the performance of the manipulators.

Figure 2 shows that there are three main groups of micromanipulators. The first group comprises the ultrahigh precision nanomanipulators with highly limited working space. For example the 6-PSSS compliant mechanism discussed in (Tung-Li Wu et al. 2008) provides an $8 \times 8 \times 8\ \mu m^3$ workspace with a resolution of 5 nm. The (major) second group represents micromanipulators with high accuracy and a workspace in the millimeter range. The range for the minimum workspace and accuracy are shown in Fig. 2 (gray area). The Agietron micronano is a sample of this group. With a $10 \times 10 \times 10\ \text{mm}^3$ workspace it can position objects with an accuracy of 0.16 μm in average (Lubrano and Clavel 2010). The third group is represented, for example, by the TRIGLIDE^S. It has a $112 \times 112 \times 112\ \text{mm}^3$ workspace with an average repeatability of 0.3 μm. This group comprises precision manipulators with an increased range of motion. The large workspace in case of the TRIGLIDE^S is achieved by using shape memory alloy to increase the range of motion of the flexure hinges themselves up to $\pm 30°$ (up to tenfold increase) (Raatz 2006).

4 Conclusions

In this paper, a comprehensive classification for robots was developed which is also valid for flexure hinge based micromanipulators. Different design and kinematic attributes were identified and discussed for micromanipulators. The developed classification was applied to list existing micromanipulators. The performance comparison of different micromanipulators has shown that three groups of micromanipulators can be distinguished: (1) ultrahigh precision nanomanipulators with highly limited workspace, (2) micromanipulators with high accuracy and a workspace in millimeter range and (3) precision manipulators with an increased range of motion. A systematical analysis of suitable structural designs has to be performed to enable new developments of flexure hinge based micromanipulators. Furthermore, a stiffness analysis of suitable structural designs is required to ensure high precision.

Acknowledgments The research work reported here was possible within a joint research project, which also includes the Laboratory for Machine Tools and Production Engineering and the Facility for Electron Microscopy of RWTH Aachen and is funded by the German Grant Authority DFG.

References

Beltrami I, Joseph C, Clavel R, Bacher JP, Bottinelli S (2004) Micro- and nanoelectric-discharge machining. J Mater Process Technol 149(1–3):263–265. doi:10.1016/j.jmatprotec.2004.03.002

Bleicher F (2003) Parallelkinematische Werkzeugmaschinen, NWV Technik. NWV Neuer Wissenschaftlicher Verlag, Wien

Cao W, Ding H, Zi B, Che Z (2013) New structural representation and digital-analysis platform for symmetrical parallel mechanisms. Int J Adv Robot Syst 10:1. doi:10.5772/56380

Corves B, Müller R (2013) Kinematik, Dynamik und Anwendungen in der Robotik. Lecture notes. RWTH Aachen University

Fazenda N, Lubrano E, Rossopoulos S, Clavel R (2006) Calibration of the 6 DOF high-precision flexure parallel robot "Sigma 6". In: Neugebauer R (ed) Parallel kinematic machines in research and practice [proceedings], vol 33, Reports from the IWU. Verl. Wiss. Scripten, Zwickau, pp 379–398

Garg V, Nokleby SB, Carretero JA (2009) Wrench capability analysis of redundantly actuated spatial parallel manipulators. Mech Mach Theory 44(5):1070–1081. doi:10.1016/j.mechmachtheory.2008.05.011

Haun M (2007) Handbuch Robotik. Programmieren und Einsatz intelligenter Roboter. Springer, Berlin/Heidelberg/New York

Howell LL (2001) Compliant mechanisms. Wiley, New York

Howell LL, Magleby SP, Olsen BM (eds) (2013) Handbook of compliant mechanisms. Wiley, Chichester

Lu TF, Handley DC, Yong YK, Eales C (2004) A three-DOF compliant micromotion stage with flexure hinges. Ind Robot Int J 31(4):355–361. doi:10.1108/01439910410541873

Lubrano E, Clavel R (2010) Compensation of thermal effects and cutting-forces acting on ultra high-precision robots. In: Borgmann H (ed) Conference proceedings/Actuator 10, 12th international conference on new actuators & 6th international exhibition on smart actuators and drive systems, Bremen, 14–16 June 2010. WFB, Div. Messe Bremen, Bremen, pp 391–396

Naval M (1989) Roboter-Praxis. Aufbau, Funktion und Einsatz von Industrierobotern, 1st edn, Vogel-Fachbuch Technik. Vogel, Würzburg

Pham HH, Chen IM (2006) Micro-manipulator design based on selectively actuated flexure parallel mechanisms. In: 2006 I.E. conference on robotics, automation and mechatronics, Bangkok, pp 1–8

Pham P, Regamey YJ, Fracheboud M, Clavel R (2005) Orion MinAngle: a flexure-based, double-tilting parallel kinematics for ultra-high precision applications requiring high angles of rotation. Tokyo, ISR Proceedings, 2005. In: 36th international symposium on robotics, Tokyo

Pierrot F (2002) Parallel mechanisms and redundancy. In: 1st International Colloquium, Collaborative Research Centre 562, Braunschweig, pp 261–277

Raatz A (2006) Stoffschlüssige Gelenke aus pseudo-elastischen Formgedächtnislegierungen in Pararellrobotern, Schriftenreihe des Instituts für Werkzeugmaschinen und Fertigungstechnik der TU Braunschweig. Vulkan-Verl, Essen

Röse A (2011) Parallelkinematische Mechanismen zum intrakorporalen Einsatz in der laparoskopischen Chirurgie. Inst. für Elektromechanische Konstruktionen, Darmstadt

Shuo Hung Chang, Chung Kai Tseng, Hon Chan Chien (1999) An ultra-precision XYΘZ piezo-micropositioner Part II: experiment and performance. IEEE Trans Ultrason Ferroelect Freq Contr 46(4):906–912. doi:10.1109/58.775657

Tang X, Chen IM, Yang G (2006) Nonlinear modeling method of a large-displacement and decoupled XYZ flexure parallel mechanism. In: 2006 9th international conference on control, automation, robotics and vision, Singapore, pp 1–6

Tung-Li Wu, Jia-Hao Chen, Shuo-Hung Chang (2008) A six-DOF prismatic-spherical-spherical parallel compliant nanopositioner. IEEE Trans Ultrason Ferroelect Freq Contr 55 (12):2544–2551. doi:10.1109/TUFFC.2008.970

Yang G, Teo TJ, Chen IM, Lin W (2011) Analysis and design of a 3-DOF flexure-based zero-torsion parallel manipulator for nano-alignment applications. In: 2011 I.E. international conference on robotics and automation (ICRA), Shanghai, pp 2751–2756

Axis Cross-Coupling Reduction on a High Bandwidth XY Flexure Stage

A. Ruiz, F.J. Campa, O. Altuzarra, V. Petuya, C. Pinto, and A. Hernández

Abstract Flexure stages are widely used in the field of micro and nanopositioning because of their repeatability and precision, which is due to the lack of joints that induces uncertainty. In the present Twork, a XY flexure stage has been analyzed and redesigned to reduce the axis cross-coupling, which is due to a parasitic motion in the amplification stage of the mechanism. The performed modifications result in a high reduction of the undesired motion, while maintaining a similar amplification and natural frequencies. The main advantage of the redesign is that the cross-coupling reduction is achieved mechanically, instead of using the position control loops.

Keywords Micro/Nanopositioning • Flexure stage • Compliant mechanisms • Mechatronics

1 Introduction

At present, flexure nanopositioning stages actuated by piezoelectrics are devices that have found many high technological applications in advanced scientifical instrumentation, microscopics, litography, nanometrology, optic systems and assembly of nanostructures (Kim and Choi 2012). Through the controlled deformation in several points of their monolithic structure, these compliant mechanisms allow generating motions with great repeatability, precision, speed and a high bandwidth. Those flexible points are flexure joints, that substitute the traditional joints used in typical mechanisms overcoming the problems they usually suffer, as friction, clearances, hysteresis and wear (Yong et al. 2012). They are usually manufactured using wire electro-discharge machining, where the grooves define the profile of the compliant mechanism.

Furthermore, these compliant mechanisms can be designed with a serial kinematic chain or a parallel kinematic chain to achieve several degrees of freedom (dof) in the end effector or platform of the mechanism. There are developments to

A. Ruiz (✉) • F.J. Campa • O. Altuzarra • V. Petuya • C. Pinto • A. Hernández
Department of Mechanical Engineering, University of the Basque Country UPV/EHU, Bilbao, Spain
e-mail: antonio.ruiz@ehu.es; fran.campa@ehu.es; oscar.altuzarra@ehu.es; victor.petuya@ehu.es; charles.pinto@ehu.es; a.hernandez@ehu.es

E.-C. Lovasz et al. (eds.), *Microactuators and Micromechanisms*, Mechanisms and Machine Science 30, DOI 10.1007/978-3-319-15862-4_6

generate motions with 1 dof (Kim and Choi 2012), 2 dof (Wang et al. 2011; Yong et al. 2009), 3 dof (Kim et al. 2010; Hassani and Tjahjowidodo 2013; Kenton and Leang 2012), or 6 dof (Liang et al. 2011; Dunning et al. 2013). In their design, there are usually two different parts: an amplification stage, responsible for amplificating the deformation of the piezoelectrics, and the guiding stage, which guides the platform or end effector of the mechanism along the directions defined by the degrees of freedom of the mechanism (Kim et al. 2010).

Nevertheless, the main drawbacks of these mechanisms are a more complex design, because we have to control the elastic deformation of the joints, and the appearance of parasitic motions. Due to their geometrical configuration this undesired motion is a coupled motion. These motions can be compensated by control, although it is always better to have an optimized mechanical design as a starting point. To achieve this, there are several rules, as the use of symmetric configurations or parasitic motion cancellation mechanisms (Kim et al. 2010). In the present paper, some of these rules will be applied to a high bandwidth nanopositioning stage to decrease the axis cross-coupling.

2 Description of the XY Flexure Stage Mechanism

Although the dimensions are not exactly the same the initial design of the stage studied within this paper is extracted from the work of Yong et al. (2009), see Fig. 1. This stage mechanism consists of three main parts.

First, the outer section provides an amplification of the piezoelectric displacements, s1 and s2, by means of an amplification lever and circular flexures. This type of flexure is selected because it provides more accurate rotational motions. With the length of the levers, the motion of the two piezoelectrics can be amplified directly to the X and Y axes, see Fig. 2.

The second part is the inner section, which is the XY stage itself. Due to the symmetric arrangement of its thin webs it generates the XY motion of the platform.

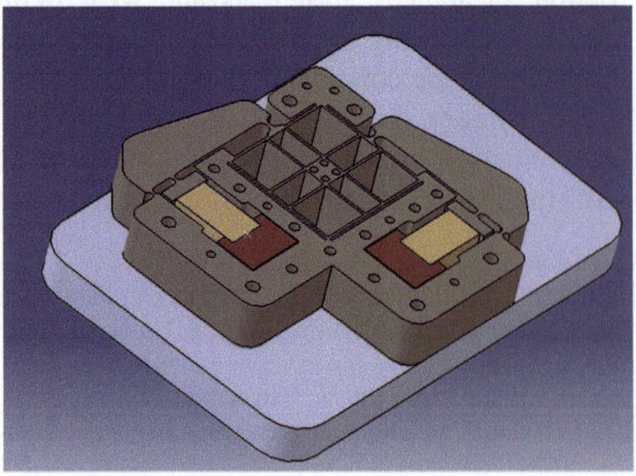

Fig. 1 Studied XY nanopositioning stage

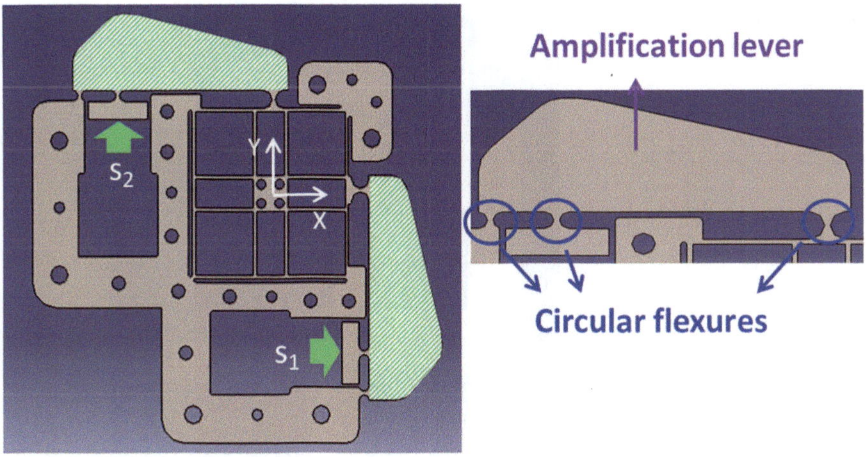

Fig. 2 Outer section: amplification stage

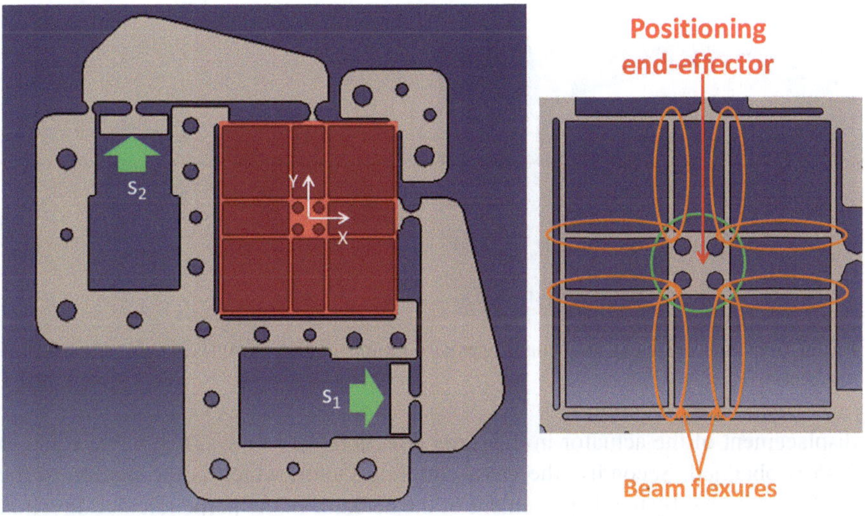

Fig. 3 Inner section: XY stage

This part of the stage is initially designed to minimize the coupling between the X and Y motions. For that reason, it has four pairs of beam flexures that they are rigid along the direction of the motion but flexible in the axis perpendicular to the motion, see Fig. 3.

Finally, the third part is the fixed frame, which is fixed to the base surface not allowing any type of movements in this part of the structure, see Fig. 4.

To analyze the performance of the stage, ANSYS has been used to perform several analyses, obtaining the main features of the XY flexure stage. Firstly, the amplification factor, which is the motion in the platform divided by the

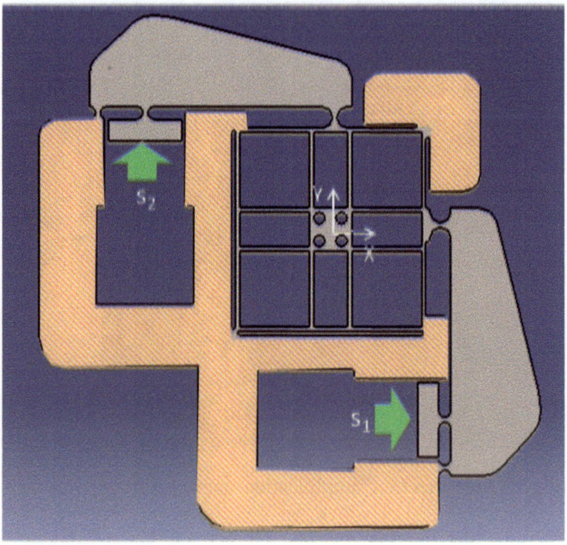

Fig. 4 Fixed frame

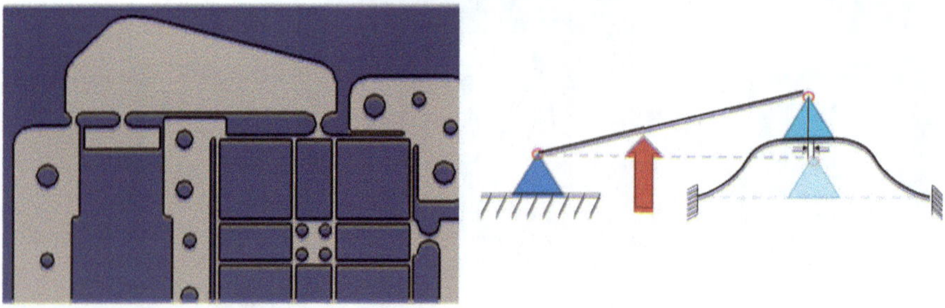

Fig. 5 XY axes cross-coupling main reason: parasitic motion in a lever type amplification mechanism

displacement of the actuator in the same direction (x/s1 or y/s2). A value equal to 3.45 is obtained. Secondly, the cross-coupling factor, which is the motion in the platform divided by the displacement of the actuator in the perpendicular direction (x/s2 or y/s1). A factor equal to 0.035 is obtained.

Also, the modal analysis reveals that the first resonance frequency appears at 1,769 Hz.

An aluminum alloy 7,075 with a Young modulus E = 72 GPa, a Poisson's ratio of 0.33 and a density of 2.81 g/cm^3 was assigned to the stage to perform the different analyses.

Although the reached amplification and the first natural frequency are satisfactory, the axis cross-coupling in open-loop control is too high for the precision requirements that these mechanisms must fulfil.

This undesired motion, which is represented on the right side of Fig. 5, happens mainly because of the rotation of the lever in the amplification stage and is a typical drawback of this kind of amplification mechanisms.

3 Mechanical Redesign to Minimize Cross-Coupling

In order to minimize the cross-coupling between X and Y axis, the following changes were made to the initial design.

First, instead of connecting directly the end of the lever arm to the second stage, an intermediate beam connected by two circular flexure hinges at its ends was added. The result is that the parasitic motion of the lever is somewhat absorbed by the rotation of the intermediate beam, see Fig. 6.

Second, to reduce the parasitic motion even more, a new layout with two symmetrical lever arms and intermediate beams has been designed. As it is shown on the right side of Fig. 7, the benefit of this layout is the total cancellation of the parasitic motion of the lever arms. However, the design is more complex as there is a new translation component pulling the lever arms at the same time, whose deflection must be controlled in order to propagate the motion correctly.

To obtain the dimensions of the elements in our stage, we have treated our problem as planar kinematics one, where the velocities can be assimilated to the

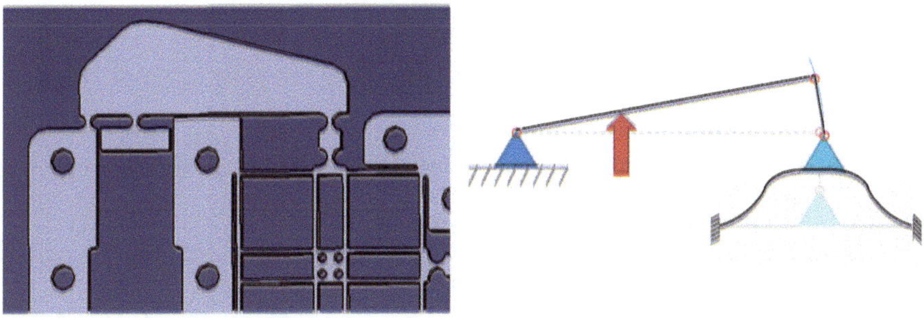

Fig. 6 Redesigned amplification stage with intermediate beam

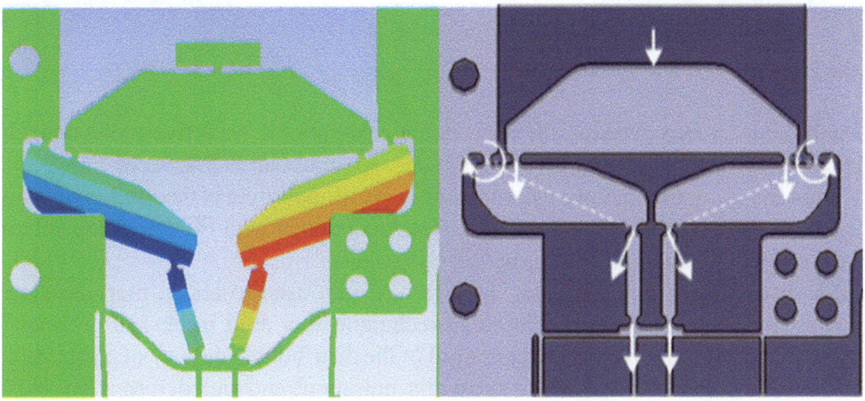

Fig. 7 Redesigned amplification stage with parasitic motion cancellation based on twin levers: (*Left*) Horizontal displacement. (*Right*) Cancellation mechanism

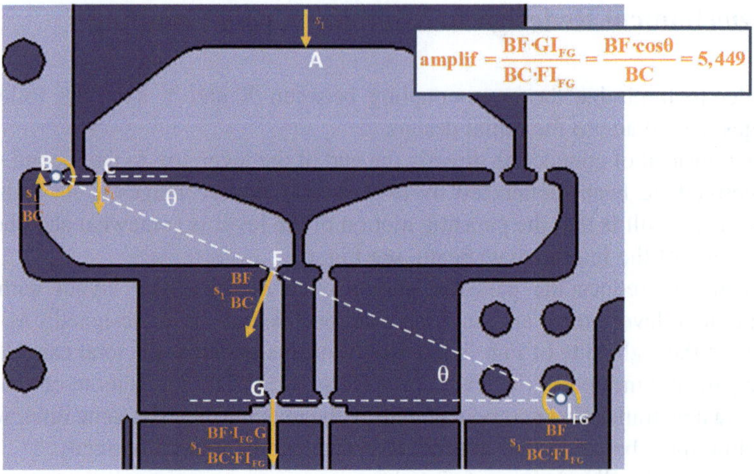

Fig. 8 Planar kinematics problem: obtaining of dimensions

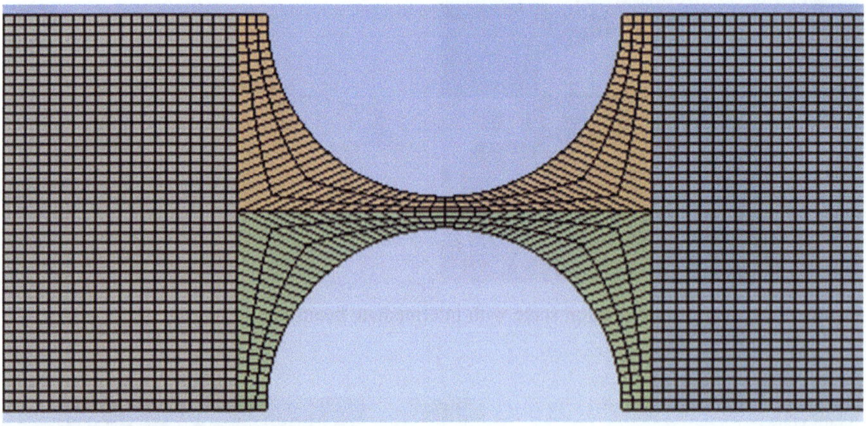

Fig. 9 Mapped mesh of circular flexures

small deformations. Making the appropriate calculations, we arrive to the conclusion that the maximum amplification which we can obtain is 5,449, see Fig. 8.

Finally, an optimization of the thickness of all the circular flexure hinges was made. To perform this, we compare our results obtained with FEM with the values from analytical equations and experimental results.

A special mesh in all circular flexures has been used to obtain more accurate results, see Fig. 9. A mapped meshing technique was used instead of a "smart" meshing because it provides better control of the distribution and size of elements in an area. It is useful in areas where stress concentrations and high deformations exist (Yong et al. 2008).

Table 1 Comparison of results (joint R = 3 mm; t = 0.5 mm; b = 12.7 mm; aluminum 7,075)

	Experimental[a]	Theoretical	FEM
K_T (Nm/rad)	6.289	6.682	6.695
K_Y (N/μm)	0.677	0.659	0.89
K_X (N/μm)	–	194.381	166.12

[a]Experimental values obtained from the article "Comparison of circular flexure hinge design equations and the derivation of empirical stiffness formulations"

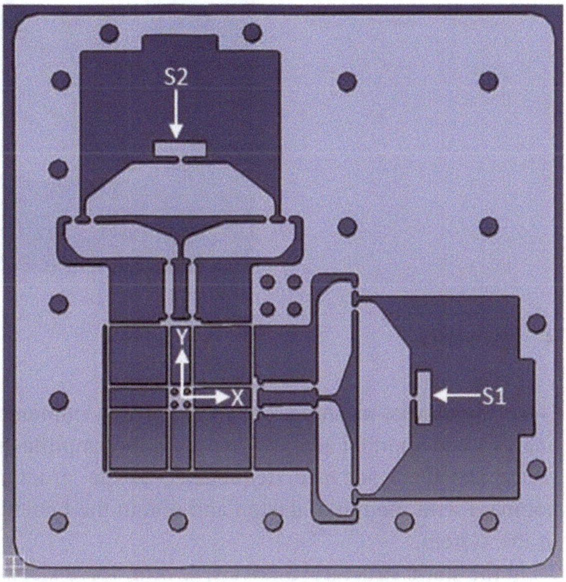

Fig. 10 Redesigned amplification stage with parasitic motion cancellation based on twin levers

In the Table 1, the similarity between the results obtained with the different methods for a specific case, can be observed.

In our work, a range of thicknesses for a radius of 0.65 mm was studied in order to reach an acceptable compromise between the amplification achieved and the stiffness of the stage. Circular flexure hinge design equations (Yong et al. 2008) were used to study the compliance of the flexures for the whole $0.05 < t/R < 0.65$ range (R is the radius and t is the neck thickness). Once all the results were analyzed, it was concluded that the optimum dimensions for the circular flexure hinges to fulfill our requirements were: radius equal to 0.65 mm and thickness equal to 0.3 mm, with a structural thickness in the mechanism equal to 12.7 mm.

Applying all these changes in the initial design, we arrive to the following resulting design, see Fig. 10.

In the Fig. 11, we can observe the displacement in the Y direction in our redesigned stage.

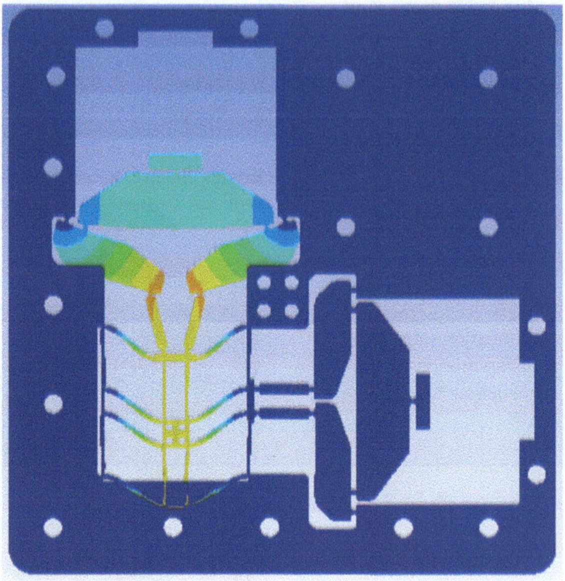

Fig. 11 Redesigned stage: displacement in the Y direction

4 Results

Once the correct meshing was made, it was implemented again a static structural analysis and a modal analysis to obtain the amplification factor, the cross-coupling factor and the resonance frequencies of the structure. The purpose of this is to compare with the initial design and obtain the benefits than the redesign introduce in the system.

In the static structural analysis, several force steps in the range of [0–400] N are applied to the face where the piezoelectric actuator acts. Then the displacement of this face and in the center of the platform in X and Y directions were measured. Thus, it was possible to calculate the axis cross-coupling and the amplification factor.

With our new XY flexure stage, the parasitic motion suffers and important reduction, around a 90 % from the initial stage. The benefits of using our redesign to reduce the cross-coupling are noticeable, as can be seen in Fig. 12, where is represented the parasitic motion in the platform against the displacement of the actuator.

Comparing the desired motion between the initial design and the new redesign, we can observe that the amplification in the second case is smaller than in the initial structure, see Fig. 13, where is represented the desired motion in the platform against the displacement of the actuator. The main cause is the shorter length of the lever in the redesigned stage, in order to maintain a similar size between the initial and the new mechanism.

After performing the modal analysis, it can be observed that the frequencies appear in pairs, due to the symmetry of the structure in the XY directions, see

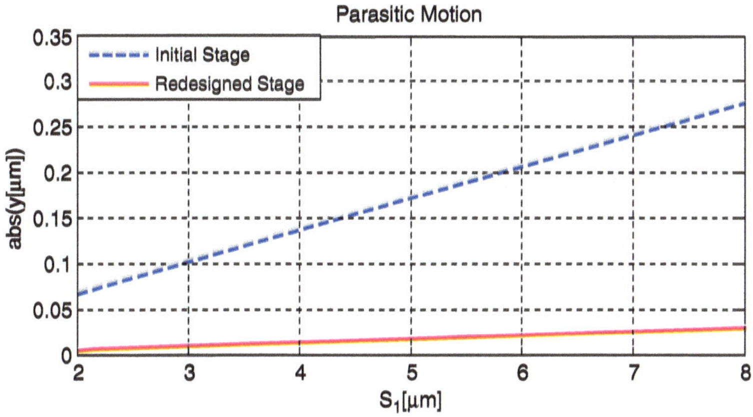

Fig. 12 Comparison of the axes cross-coupling of the stages

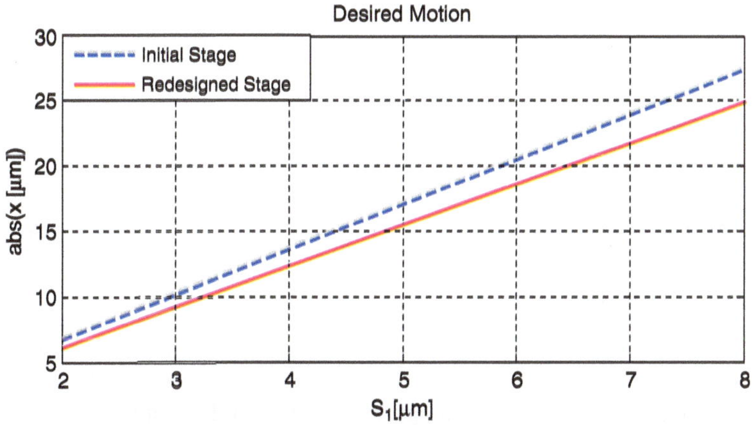

Fig. 13 Comparison of the amplification of the stages

Fig. 14, where is represented the first and the second resonance frequencies of each flexure stage. It is estimated that the obtained value for the first resonance frequency of 1,522 Hz will allow reaching the desired high bandwidth, that is, in the range of 100–300 Hz in a closed-loop system.

To sum up the results obtained in the different analysis the main features of the different stages are shown in the Table 2.

As it can be seen in Table 2, a similar amplification and first resonance frequency are achieved with the new design while decreasing the cross-coupling by a factor of 10, which was the main objective of this work to fulfill the requirements of accuracy of this type of stages.

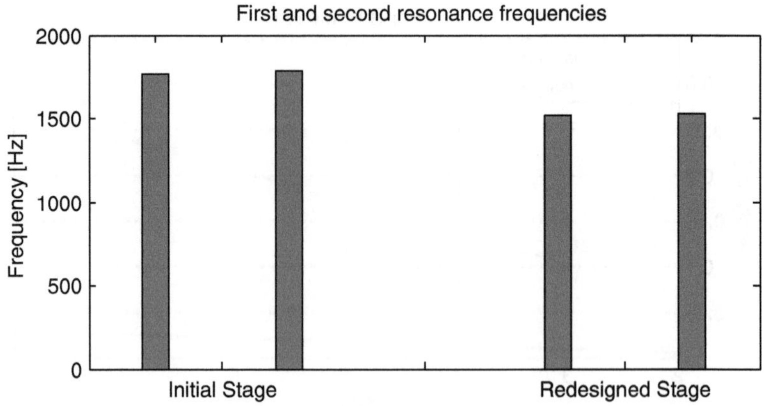

Fig. 14 Estimated frequency responses of the stage

Table 2 Results obtained for each stage

	Initial stage	Redesigned stage
Amplification factor (x/s_1; y/s_2)	3.45	3.13
Cross-coupling factor (y/s_1; x/s_2)	0.0349	0.004
1st natural frequency [Hz]	1,769	1,522

5 Conclusions

The main contributions of the present paper are:

- A flexure XY stage for high bandwidth nanopositioning has been analyzed and the causes of its high axis cross-coupling have been determined: the parasitic motion of the lever arm in the amplification stage induces an undesired motion of the guiding mechanism.
- A redesign of the initial mechanism has been done, focusing on three main aspects:
  1. Addition of a new beam that links the lever arm and the guiding mechanism, reducing the pulling force of the lever onto the guiding mechanism and absorbing part of the parasitic movement.
  2. Symmetrical layout, with two lever arms and linking beams for each axis that cancels the parasitic motions of the amplification stage.
  3. New dimensions for the flexure joints, finding a compromise between the amplification reached and the stiffness, which limits the first modal frequency and, thus, the bandwidth of the mechanism.

- The static and modal analysis performed by Fem of the redesign shows a similar amplification and first natural frequency, but reduces the axis cross-coupling by a factor of 10 from the initial value. The straightness, cross-coupling divided by amplification, is reduced from a 1.012 % of the desired motion to a 0.128 %.
- This cross-coupling reduction is achieved mechanically instead of using the control

Further work will consist on improving our stage with new changes in the structure and manufacturing and testing a prototype, to validate the promising results shown by the performed analysis.

Acknowledgments The authors of this paper wish to acknowledge the finance received from the Spanish Government via the Ministerio de Educacion y Ciencia (Project DPI2011-22955), the ERDF of the European Union, the Government of the Basque Country (Project GIC07/78, IT445-10), and the University of the Basque Country (Project EHUA13/30 and ZABALDUZ program). Thanks are also addressed to Mikel Abasolo for his assessments regarding the use of ANSYS and meshing and to EGILE Corporation XXI for their high valuable recommendations.

References

Dunning AG, Tolou N, Herder JL (2013) Compact low-stiffness six degrees of freedom compliant precision stage. Precis Eng 37:380–388

Hassani V, Tjahjowidodo T (2013) Dynamic modeling of 3-DOF pyramidal-shaped piezo-driven mechanism. Mech Mach Theory 70:22–245

Kenton BJ, Leang KK (2012) Design and control of a three-axis serial-kinematic high-bandwidth nanopositioner. IEEE/ASME Trans Mech 17(2):356–369

Kim J, Choi Y (2012) A millimeter-range flexure-based nano-positioning stage using a self-guided displacement amplification mechanism. Mech Mach Theory 50:109–120

Kim H, Ahn D, Chun B, Gweon D (2010) Development and optimization of a novel 3-DOF precision flexure stage. In: Proceeding of the IEEE international conference on nanotechnology joint symposium with Nano Korea, Kintex, pp 903–906

Liang Q, Zhang D, Chi Z, Song Q, Ge Y (2011) Six-DOF micro-manipulator based on compliant parallel mechanism with integrated force sensor. Robot Comput Int Manuf 27:124–134

Wang W, Han C, Choi H (2011) 2-DOF Kinematic XY stage design based on flexure element. In: Proceeding of the IEEE international conference on mechatronics and automation, Beijing, pp 1412–1417

Yong YK, Lu T, Handley DC (2008) Review of circular flexure hinge design equations and derivation of empirical formulations. Precis Eng 32:63–70

Yong YK, Aphale S, Moheimani SOR (2009) Design, identification, and control of a flexure-based XY stage for fast nanoscale positioning. 10th IEEE Trans Nanotechnol 8(1):46–54

Yong YK, Moheimani SOR, Kenton BJ, Leang KK (2012) High-speed flexure-guided nanopositioning design and control issues. Rev Sci Instrum 83:121101

Some Structural and Kinematic Characteristics of Micro Walking Robots

Adr. Comanescu, I. Dugaesescu, and D. Comanescu

Abstract The micro-robots are generally characterized by reduced dimensions and a mono-mobile mechanical structure. This includes mechanisms acted by a single motor and consequently one input and multiple outputs. Such micro-robots have a fixed displacement strategy. Their structure may include mechanisms with bars or complex ones with gears, cam and bars. The systems may be equipped with various types of sensors such as tactile, proximity, sound and light ones.

Keywords Fixed sequence robot • Walking micro-robot • Structure modeling • Kinematic modeling • Complex micro mechanism

1 Introduction

The paper illustrates the class of micro walking robots by some biomorphic constructive solutions selected form the literatures. Their system is powered by a single micro-motor generally acting two identical mechanisms parallel connected.

The robots belong to the class of robots with a fixed sequence because the sequence of movement amplitude and legs – movement and support elements can not be changed.

The bug micro robot (Fig. 1) that reacts to touch and sound is a sixed legged system. The system is powered by a single micro-motor acting two identical mechanisms parallel connected. Each mechanism in turn causes simultaneous movement of the three legs on one side.

The crab micro robot (Fig. 2) has a similar solution. The micro-motor acts to three identical mechanisms symmetrically placed and parallel connected.

The spider type stepping micro robot is shown in Fig. 3. The solution is a mechanism with one degree of mobility having two elements for displacement and support an alternative role with a central element. This central element can rotate the entire mechanism for placing and change the system on the motion trajectory.

All these solutions are inspired by the insect anatomy.

A. Comanescu (✉) • I. Dugaesescu • D. Comanescu
Department of Mechanisms and Robots Theory, University Politehnica of Bucharest, Bucharest, Romania
e-mail: adrianacomanescu@yahoo.com; dinucomanescu@yahoo.com; ileana_d1@yahoo.com

E.-C. Lovasz et al. (eds.), *Microactuators and Micromechanisms*, Mechanisms and Machine Science 30, DOI 10.1007/978-3-319-15862-4_7

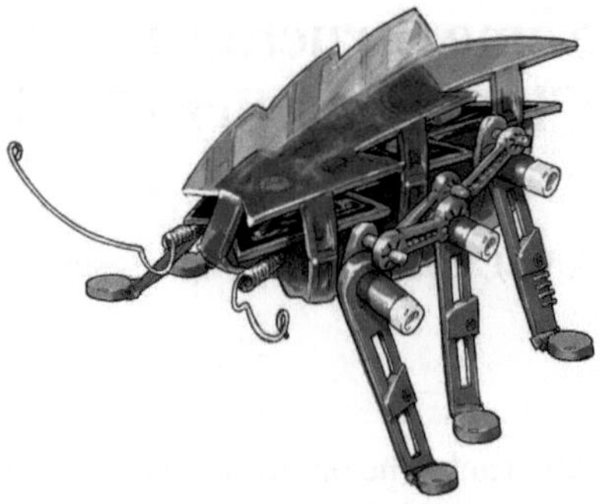

Fig. 1 The bug walking micro robot

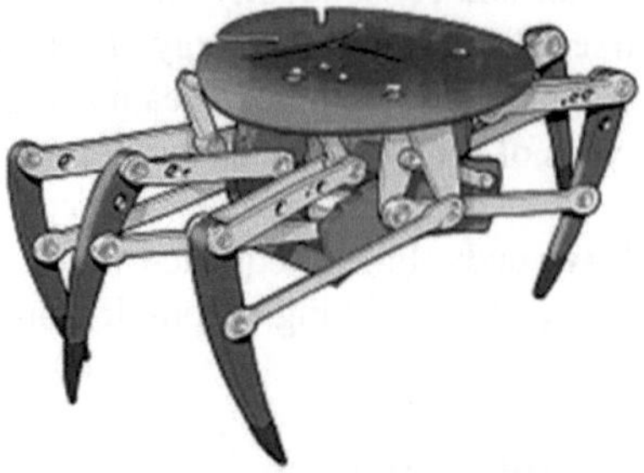

Fig. 2 The crab walking micro robot

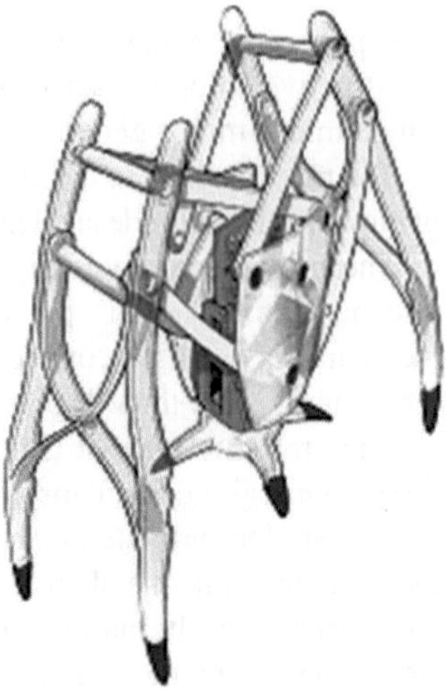

Fig. 3 The spider walking micro robot

A particular structure may be found in the construction of the micro robot presented in the Fig. 4. It is a complex mechanism with gears, cam and bars which is shown in the next part.

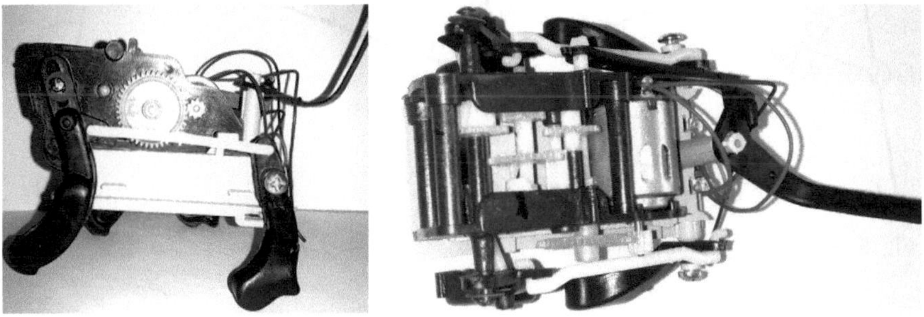

Fig. 4 A walking micro robot

2 Structural and Kinematic Particularities of the Biomorphic Walking Robots Inspired by Insects

The bug micro robot (Fig. 1) has in its structure two parallel mechanisms actuated by the same micro motor. In the Fig. 5 is given the particularities of each mechanism.

The mechanism has 7 kinematic links and 10 lower kinematic pairs relative to the platform – the bug body and its degree of mobility is given by $M = 3 \cdot 7 - 2 \cdot 10 = 1$. From structural point of view (Fig. 6) three RRR passive groups are parallel connected between the basis and the active group formed by the A active kinematic pair and the 1 link.

The geometrical constant parameters are the following: AB, BC, DC, BE, GE, GF, BF, XD, YD, XG, YG, DT3, BT4, GT7. The positional independent parameter given by the micro motor is φ_1. The dependent positional parameters are the following φ_2, φ_3, φ_4, φ_5, φ_6, φ_7 and their variation for a kinematic cycle are respectively presented by $\varphi20_k$, $\varphi30_k$, $\varphi40_k$, $\varphi50_k$, $\varphi60_k$, $\varphi70_k$ in the Fig. 7.

The trajectories of the output links extremities named T_3, T_4, T_7 are those shown in the Fig. 8.

The vertical displacement variation for these points (Fig. 9) gives the image of the sequence with which the output links are placed on the soil.

Having in view the extremities vertical displacement variation of each leg corresponding to both parallel mechanisms (Fig. 10) one may deduce the succession of each leg attaching the support surface.

The crab walking micro robot (Fig. 2) has three planar mechanisms parallel connected acted by a micro motor (Fig. 11).

Its modular structure (Fig. 12) relative to the platform is similar to the bug micro robot and has 9 links with 13 lower kinematic pairs.

The degree of mobility equal to one justifies the construction with a single motor.

The structure is based on the active modular group (A, 1) to which four passive groups RRR are connected.

The geometrical constant parameters are the following: AB, BC, DC, BE, GF, EF, XD, YD, XG, YG, ET4, BC′, D′C′, BE′, G′F′, E′F′, XD′, YD′, XG′, YG′, E′T8.

The positional independent parameter given by the micro motor is φ_1.

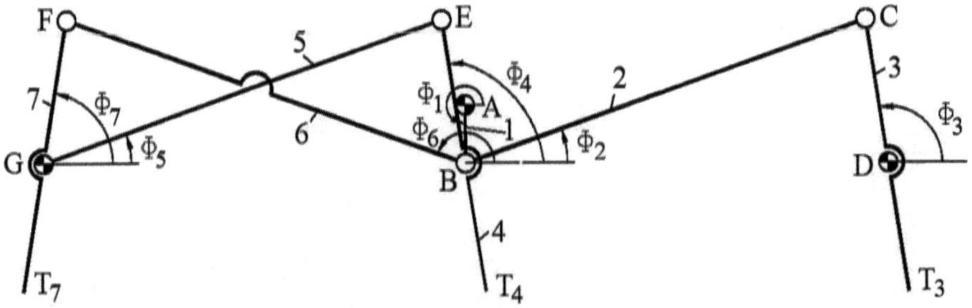

Fig. 5 The bug micro robot mechanism

Fig. 6 The structural model of the bug micro robot mechanism

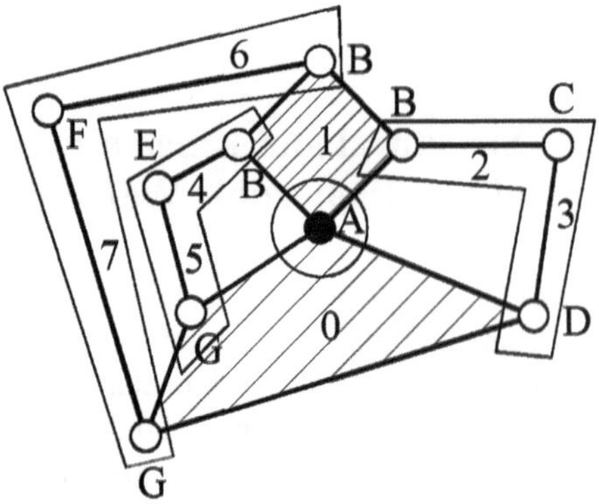

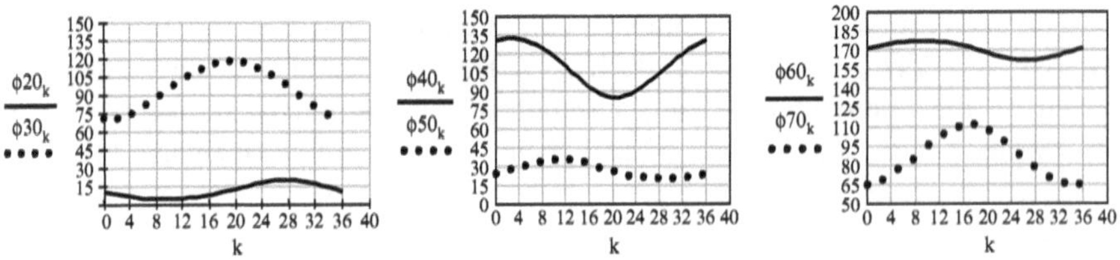

Fig. 7 The variation of the dependent positional parameters for a kinematic cycle

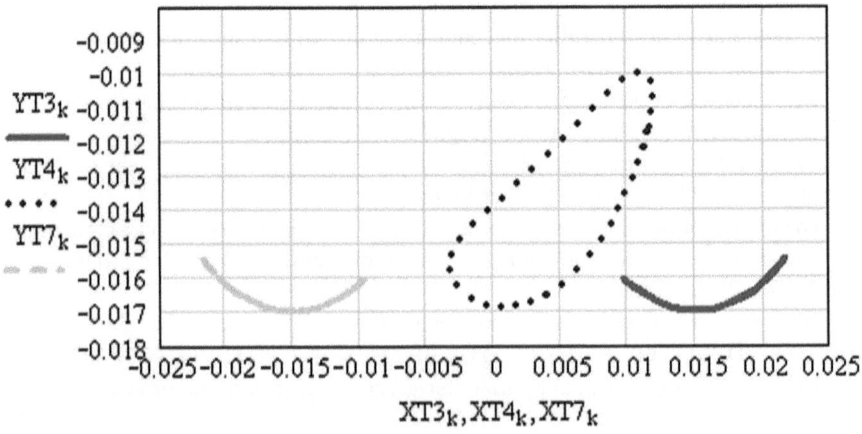

Fig. 8 The trajectories of the output links extremities

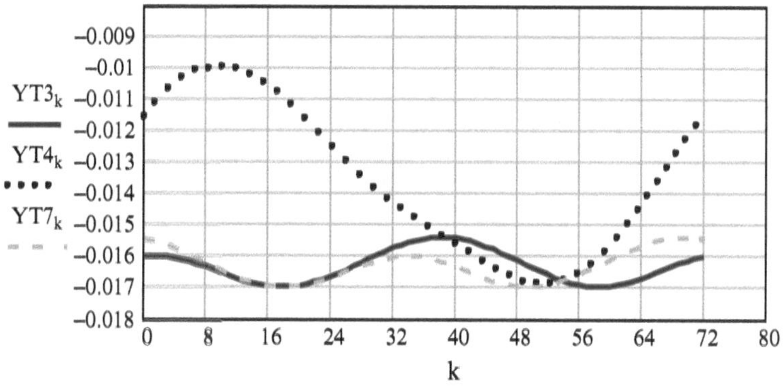

Fig. 9 The vertical displacement variation of each leg extremity

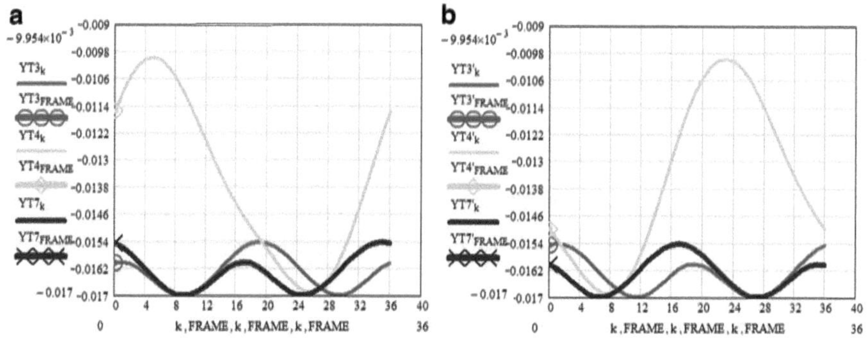

Fig. 10 The legs extremity vertical displacement. (**a**) For the left mechanism. (**b**) For the right mechanism

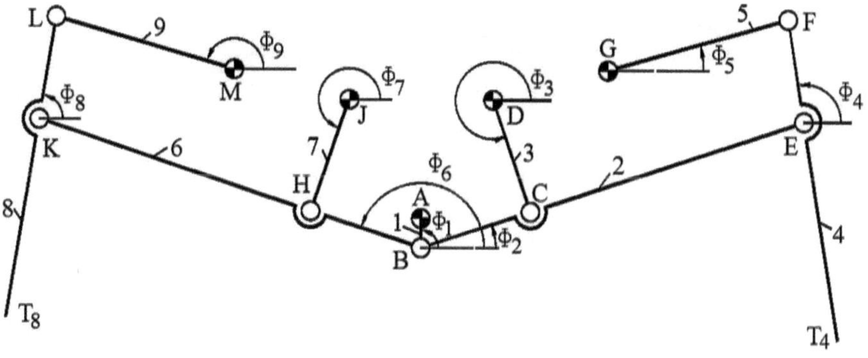

Fig. 11 The crab micro robot mechanism

The dependent positional parameters presented in Fig. 11 are the following φ_2, φ_3, φ_4, φ_5, φ_6, φ_7, φ_8, φ_9 and their variation for a kinematic cycle are respectively presented by $\varphi20_k$, $\varphi30_k$, $\varphi40_k$, $\varphi50_k$, $\varphi60_k$, $\varphi70_k$, $\varphi80_k$, $\varphi90_k$ in the Fig. 13.

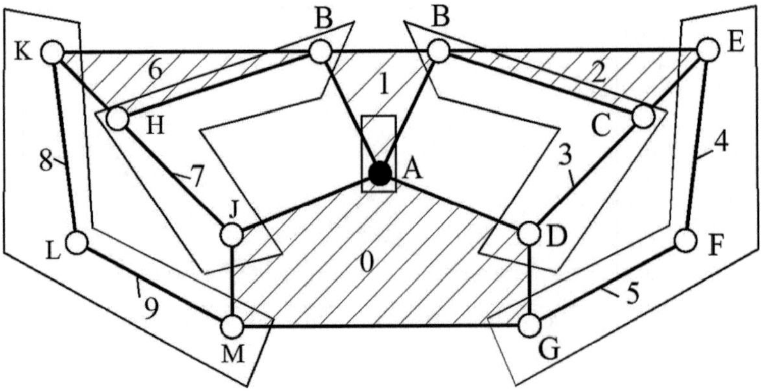

Fig. 12 The structural model of the crab micro robot mechanism

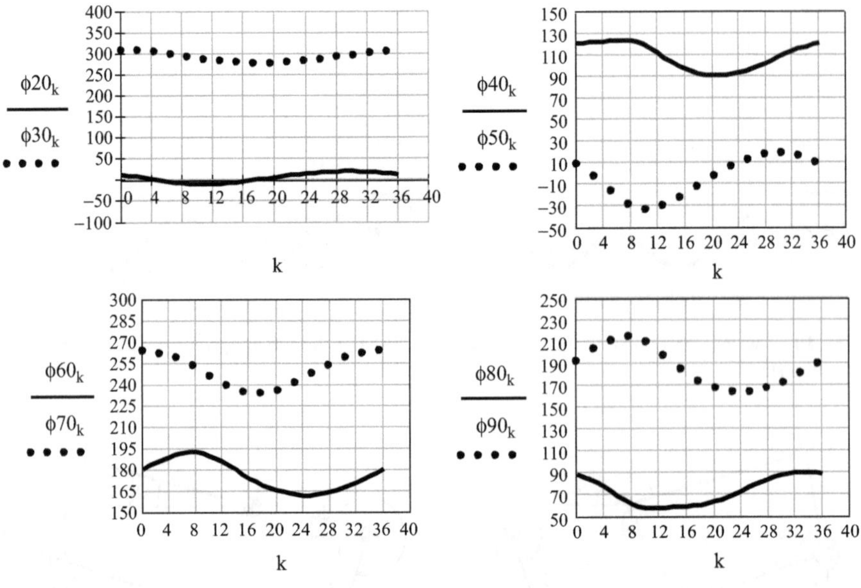

Fig. 13 The variation of the dependent positional parameters for a kinematic cycle

The trajectories of the output links extremities named T_4, T_8 are those shown in the Fig. 14.

For a kinematic cycle the 4 and 8 legs are alternatively placed on the soil. This fact is demonstrated by the vertical displacement variation which is presented in the Fig. 15.

The mechanical systems contain three mechanisms (two of them placed outside of the platform and another situated in an inner place) parallel connected (Fig. 16).

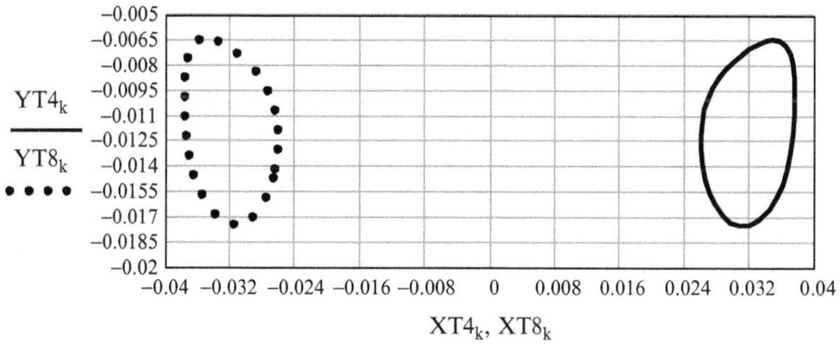

Fig. 14 The trajectories of the output links extremities

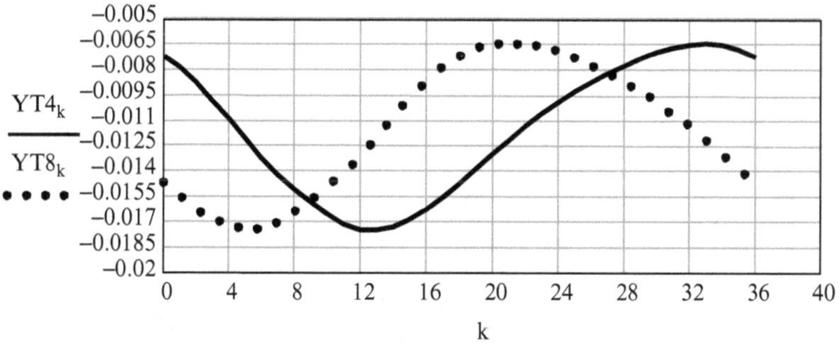

Fig. 15 The vertical displacement variation of each leg extremity for the outside mechanism

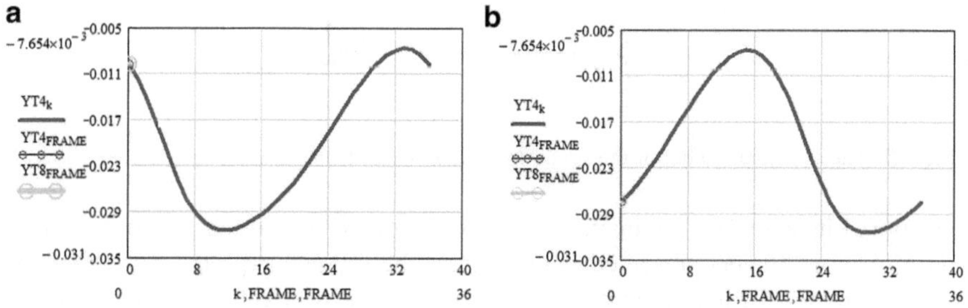

Fig. 16 The legs extremity vertical displacement for both kinds of mechanisms. (**a**) For the outside mechanism. (**b**) For the inner mechanism

The construction of the spider walking micro robot (Fig. 3) is based on the mechanism (Fig. 17) with one degree of mobility.

Its structural model (Fig. 18) includes an active modular group and four RRR passive modular groups.

The geometrical constant parameters are the following: AB, BC, GC, BD, FE, DE, DT8, BH, GH, BI, IJ, KJ, IT4, XA, YA, XG, YG, XF, YF, XK, YK, ET4. The

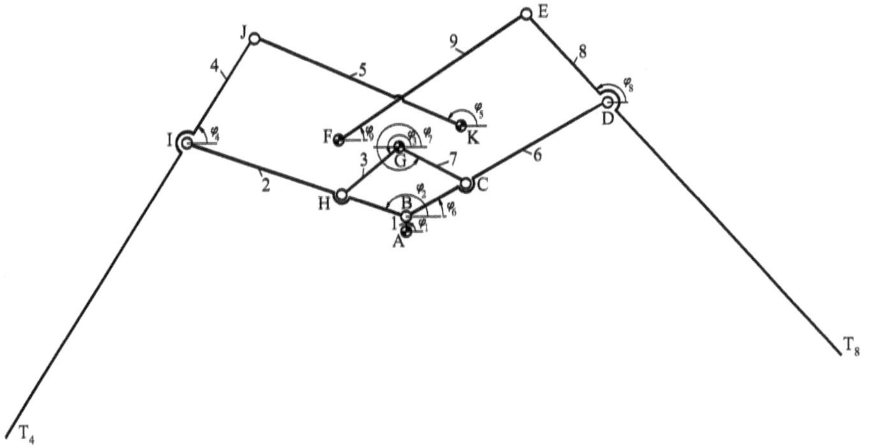

Fig. 17 The spider micro robot mechanism

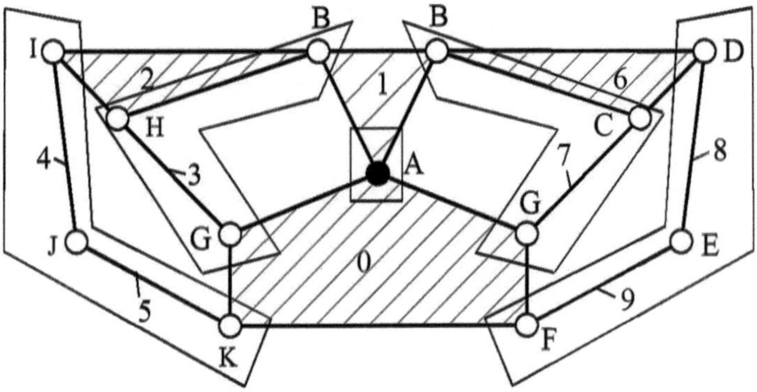

Fig. 18 The structural model of the spider micro robot mechanism

positional independent parameter given by the micro motor is φ_1. The dependent positional parameters presented in Fig. 17 are the following φ_2, φ_3, φ_4, φ_5, φ_6, φ_7, φ_8, φ_9 and their variation for a kinematic cycle are respectively presented by $\varphi20_k$, $\varphi30_k$, $\varphi40_k$, $\varphi50_k$, $\varphi60_k$, $\varphi70_k$, $\varphi80_k$, $\varphi90_k$ in the Fig. 19.

The trajectories of the leg extremities are shown in Fig. 20. They are distinct so that the displacement is achieved by the alternative leg placement on the support surface and the movement of the entire mechanism relative to the support leg. In the same time it is relevant the variation of the leg vertical displacement for an entire kinematic cycle (Fig. 20).

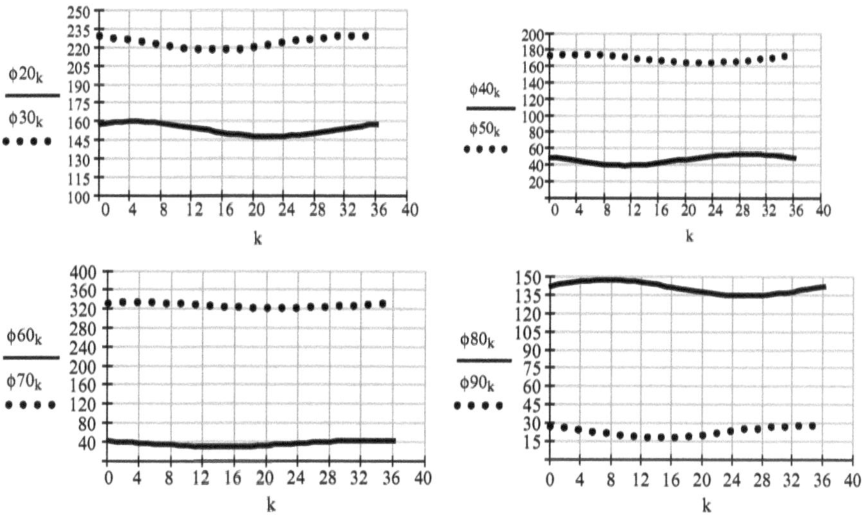

Fig. 19 The variation of the dependent positional parameters for a kinematic cycle

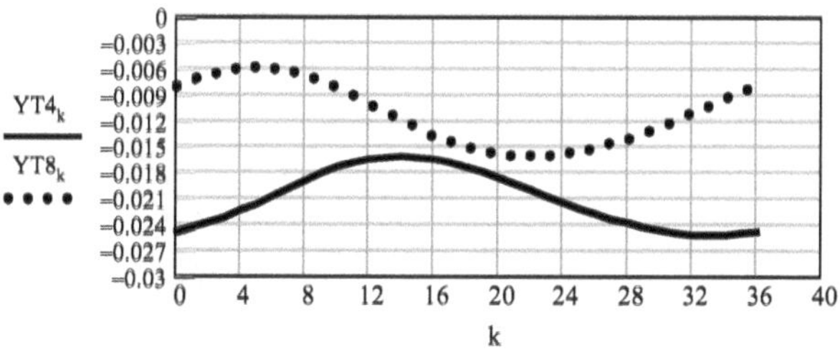

Fig. 20 The vertical displacement variation of each leg extremity

3 Particularities of the Quadruped Walking Robots

The micro robot structure has usually two or three parallel mechanisms actuated by the same micro motor. A particular structure may be found in the construction of the micro robot presented in Fig. 4. It is a complex mechanism with gears, cam and bars which is exposes in the next part.

The gear and cam complex mechanism (Fig. 21) distributes the input motion to the mechanisms acting the output links and other parts of the micro robot.

The gear mechanism (Fig. 21) has seven shafts, six of them (I, II, III, V, VI, and VII) having a rotation motion and the IV shaft a cylindrical one.

The following gears – z_1, z_{21}, z_{23}, z_{32}, z_3, z_{44}, z_4, z_{54}, z_{56}, z_6, z_7 are rigidly connected to their shaft. In the same time the gear with z_{43} teeth may have a translation motion relative to its shaft. It also has the characteristics of a cylindrical

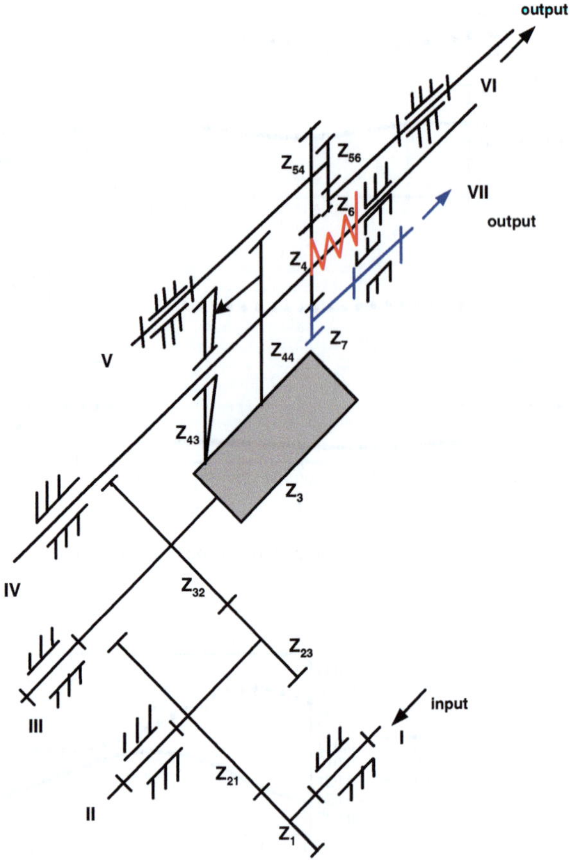

Fig. 21 The gear and cam mechanism

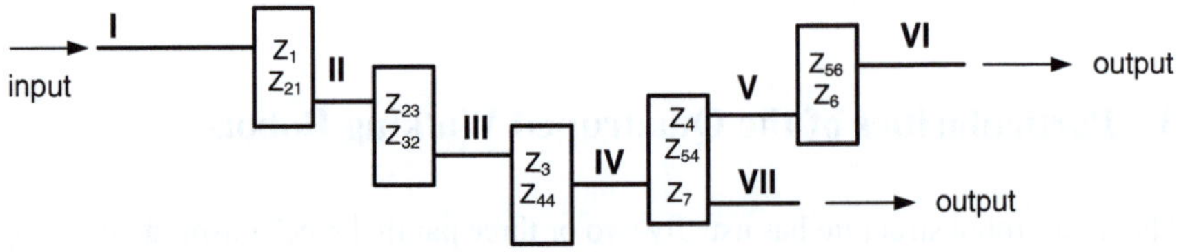

Fig. 22 The first distribution of the rotation motion

cam its cam follower being rigidly connected to the gear z_{44}. The contact between the cam and the follower is determined by a helical spring.

Due to the cam rotation motion rigidly attached to the z_{43} gear the shaft IV may have two positions.

In the first position (Fig. 22) the z_4 gear determines the rotation of the shafts V, VI and VII to the outputs being on the shafts VI and VII.

In the second position (Fig. 23) the z_4 gear directly acts the shaft VI.

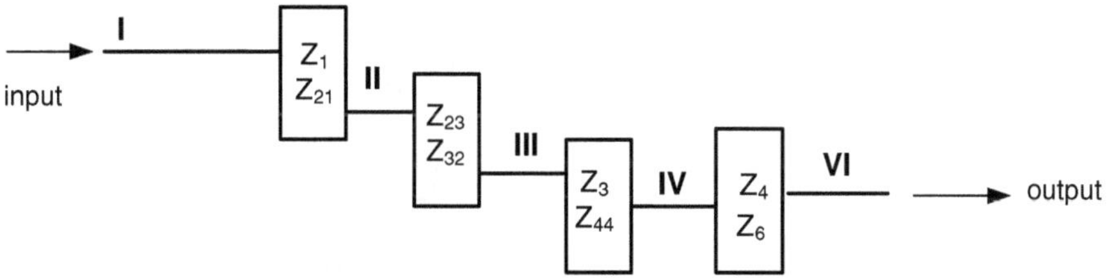

Fig. 23 The second distribution of the rotation motion

Fig. 24 The stepping micro mechanism

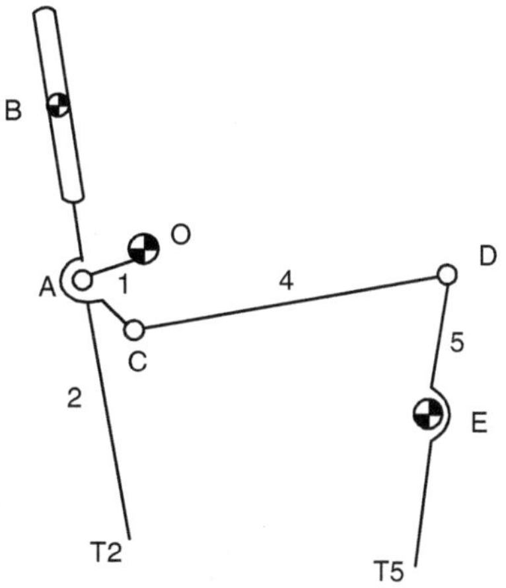

The following numbers of teeth are adopted.

z_1	z_{21}	z_{23}	z_{32}	z_3	z_{43}	z_{44}	z_4	z_{54}	z_{56}	z_6	z_7
10	42	8	16	10	31	32	22	26	16	28	22

The gear and cam complex mechanism the shaft VI (Figs. 19, 20, and 21) has in two situations different senses of rotation so that being ensured the forward and backward displacement of the micro robot.

The structural model of the stepping mechanism includes an active initial modular group (O,1) and two passive modular groups RTR and RRR (4, 5) serially connected (Comanescu et al. 2010). The micro robot structure has two parallel mechanisms actuated by the same micro motor and having a diphase of 180°.

The kinematic scheme is given in Fig. 24. The 2 and 5 kinematic links, which have the contact with the support surface in T_2 and T_5 ensure the displacement and sustain the system acted by the O active kinematic pair.

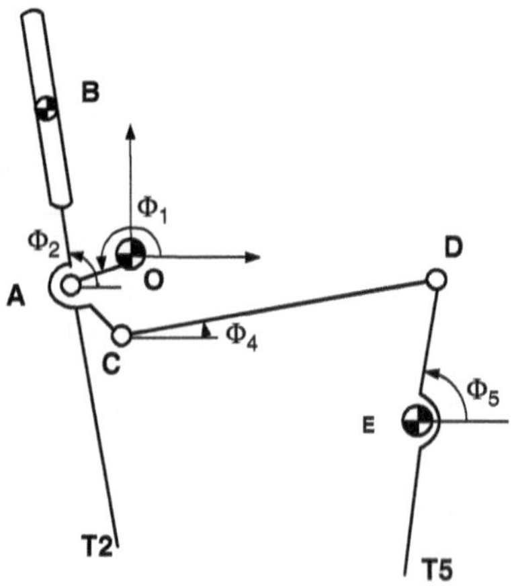

Fig. 25 The dependent positional parameters

The mechanism has 4 kinematic links and 5 lower kinematic pairs and a higher pair between the 2 link and the platform. Its degree of mobility is given by $M = 3{\cdot}4 - 2{\cdot}5 - 1 = 1$.

According to the structural models previously mentioned the positional parameters for each link are mentioned in Fig. 25. By means of standard kinematic models (Comanescu et al. 2010) for both parallel mechanisms they are determined (Fig. 26).

The legs extremity trajectories of the T_2 and T_5 points are presented in Fig. 27 for the parallel mechanisms. The initial position of each leg extreme point on the figures is marked. Thus the micro robot displacement is ensured.

In the next figure the vertical displacement variation for the leg robot extremities placed on both sides is presented (Fig. 28).

4 Conclusions

The paper brings in attention some aspects connected to the kinematic analysis of planar mechanisms applied to the micro robots. The construction of micro robots is generally based on planar mechanisms with one degree of mobility, which explains their fixed sequence of motion.

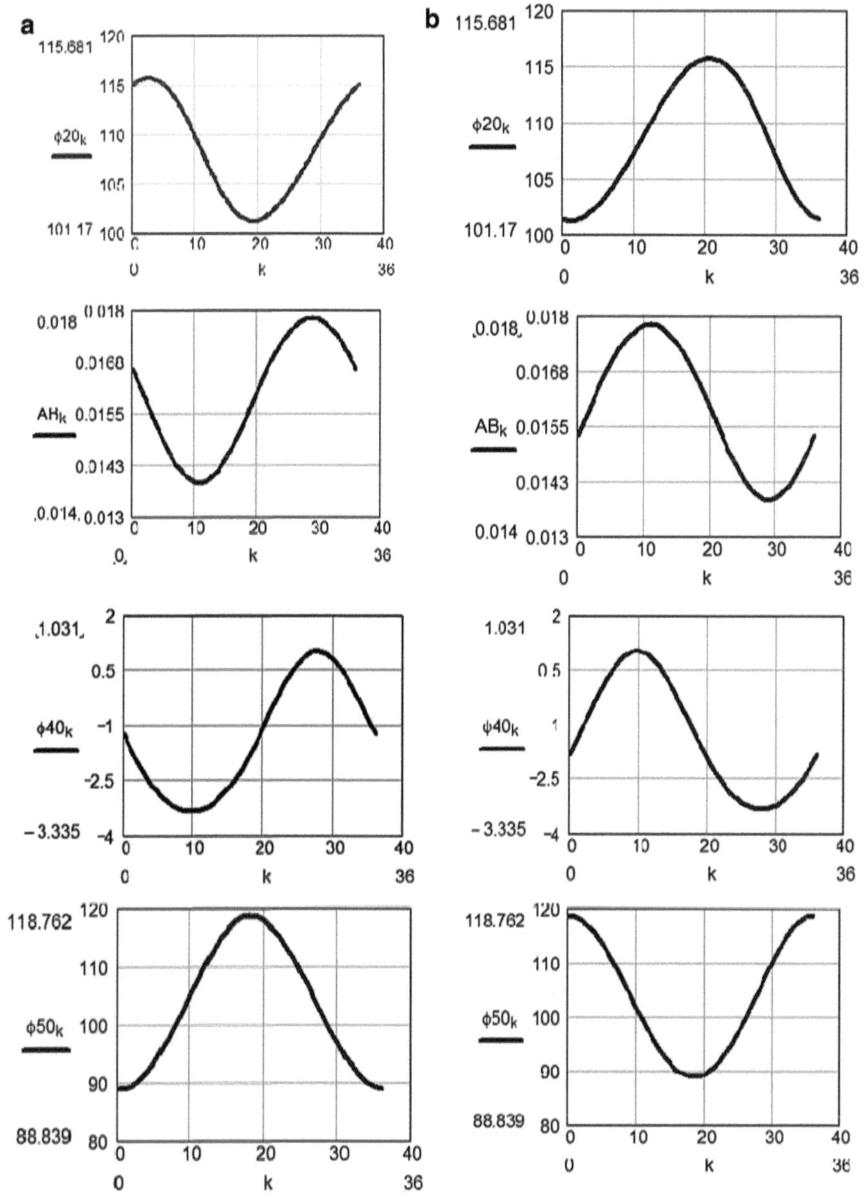

Fig. 26 The dependent parameters of both parallel stepping mechanisms. (**a**) For the left mechanism. (**b**) For the right mechanism

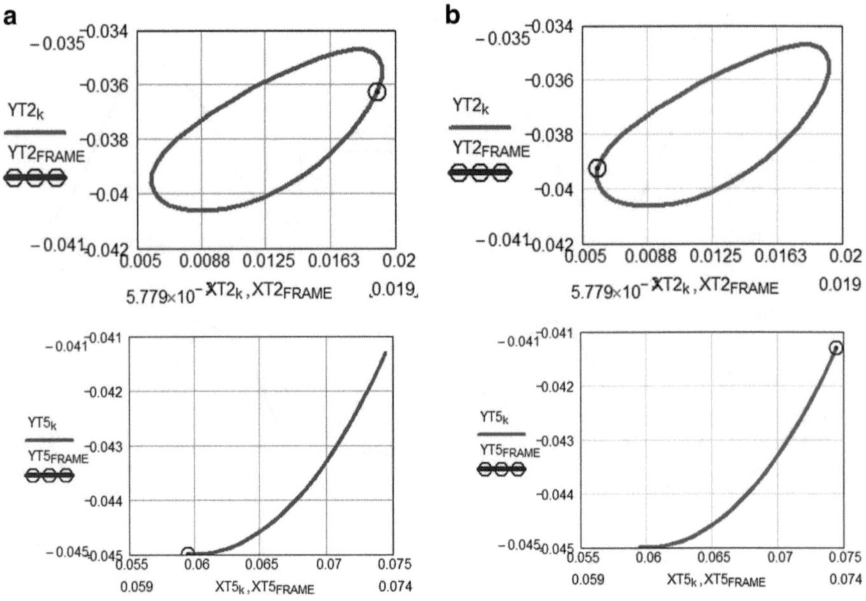

Fig. 27 The legs extremity trajectories. (**a**) For the left mechanism. (**b**) For the right mechanism

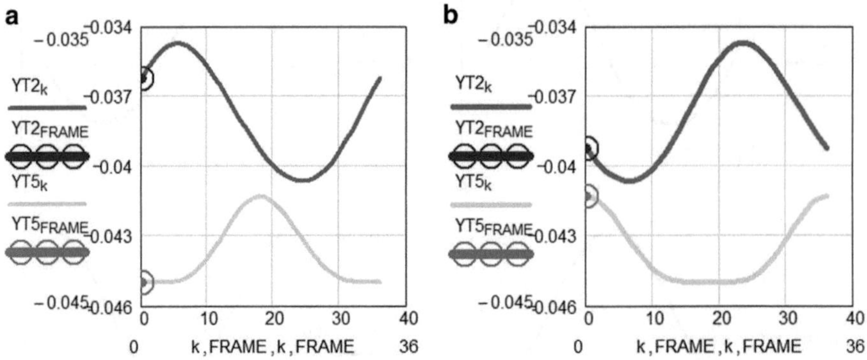

Fig. 28 The vertical displacement of the legs extremity. (**a**) For the left mechanism. (**b**) For the right mechanism

References

Angeles J (2003) Fundamentals of robotic mechanical systems: theory, methods, and algorithms, 2nd edn. Springer, New York. ISBN 978-0-387-95368-7

Comanescu A, Dugaesescu I & col. (2010) Mechanisms modelling bases. Printech, Bucharest

Comanescu A, Comanescu D, Dugaesescu I, Ungureanu LM (2013) Chapter 21: Optimal inverse models for bi-mobile mechanisms of walking robot legs. In: DAAAM international scientific book 2013. DAAAM International Publishing, Vienna, pp 417–430. doi:10.2507/daaam.scibook.2013.21. http://www.daaam.info/Downloads/Pdfs/science_books_pdfs/2013/Sc_Book_2013-021.pdf

Development of a 3-DOF Compliant Robotic Local Structure with Large Twist Angle

S. Kurtenbach, J. Siebrecht, D. Schoenen, M. Hüsing, and B. Corves

Abstract This paper presents the development of a compliant wrist joint (robotic local structure) enabling large twist angles. The result is a spherical joint with three degrees of freedom whose development is introduced starting with a general overview on suited compliant joints and the derivation of the requirements. A systematic structural synthesis and subsequently a dimensional synthesis using the pseudo-rigid-body-model is conducted. Finally, the CAD-Model is shown. The compliant robotic local structure is used in a handling system developed at the IGM facilitating a quick adaption of the kinematic structure to a new motion task through a modular and versatile concept. Several parallel kinematic robotic arms integrate an object in the kinematic structure by adapting to the object with the end-effectors.

Keywords Large twist angle • Compliant robotic local structure • Pseudo-rigid-body-model • Wrist joint • Reconfigurable handling system • Spherical linkage

1 Introduction

Now-a-days parallel handing devices are used in various application areas. The Department of Mechanisms Theory and Dynamics of Machines (IGM) at RWTH Aachen University developed a new and innovative handling system based on a reconfigurable structure with modular layout (Müller et al. 2009, 2010; Riedel et al. 2008, 2010). The handling system moves the object with six degrees of freedom in space, using a parallel kinematic structure. This results only through the integration of the object using several robotic arms. This makes it a part of the parallel kinematic structure. Figure 1 shows the entire handling system with all three robotic arms adapted to the object. A single robotic arm has six degrees of freedom; three rotational degrees of freedom in the robotic regional structure and further three in the robotic local structure. The regional structure represents a hybrid structure. A planar five-bar linkage can be rotated around a previously located main

S. Kurtenbach (✉) • J. Siebrecht • D. Schoenen • M. Hüsing • B. Corves
Institut für Getriebetechnik und Maschinendynamik, RWTH Aachen University, Aachen, Germany
e-mail: kurtenbach@igm.rwth-aachen.de; siebrecht@igm.rwth-aachen.de; schoenen@igm.rwth-aachen.de; huesing@igm.rwth-aachen.de; corves@igm.rwth-aachen.de

E.-C. Lovasz et al. (eds.), *Microactuators and Micromechanisms*, Mechanisms and Machine Science 30, DOI 10.1007/978-3-319-15862-4_8

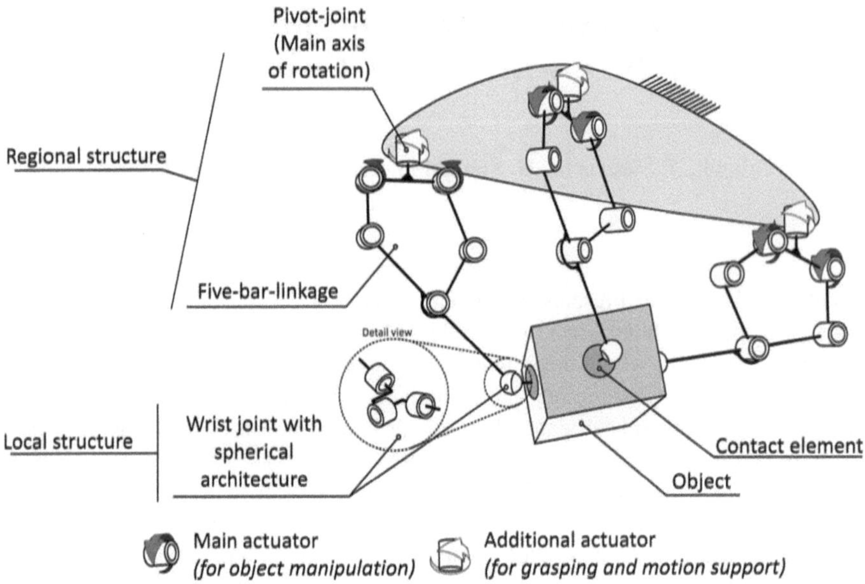

Fig. 1 PARAGRIP

axis of rotation. Drives (servo motors) are provided solely for the three degrees of freedom of the regional structure. The end-effector is operated in a purely passive form. This is to reduce the number of drives for the entire handling system to nine servo motors to move an object with three translational and three rotational degrees of freedom. Similar applications for three serial industrial robots require 18 drives.

The current local structure of the robotic arms comprises a spherical arrangement of the wrist. Figure 2a schematically shows the spherical structure with the notation of various axes. The integration of the object in the kinematic structure has been achieved in previous developments by using an electromagnet. This magnet represents the end-effector (Fig. 2b). The electrical energy is passed through the entire spherical structure by collector ring contacts, which are provided on the corresponding axes of rotation. In order to avoid a collision between the wrist joint and the object during docking, the R-axis is equipped and actuated by a stepping motor, which serves merely to pre-position the wrist joint. It is switched passive immediately after the object has been gripped.

The S-axis has a balancing and restoring spring element that holds the solenoid in the center position while not in contact with the object. A disadvantage of the current robotic local structure is the high effort for a complete housing which goes along with a limitation of the orientation workspace (small revolute angles).

In this paper, the systematic development of a compliant robotic local structure is presented with regard to the introduced handling system. Thus, a compliant robotic local structure with large twist angles and high position accuracy.

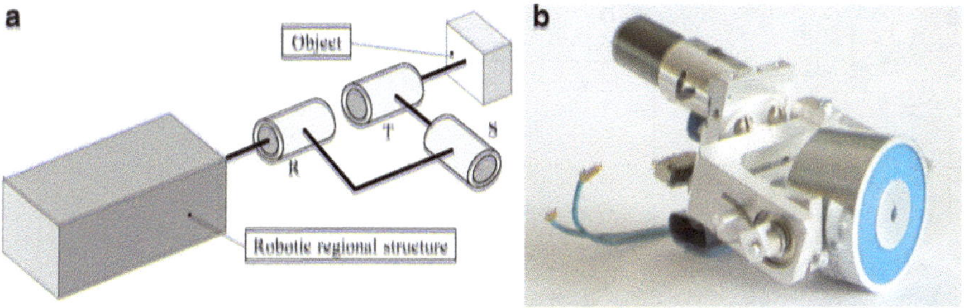

Fig. 2 Schematic representation (**a**) and image (**b**) of the current wrist joint

2 State of the Art

In the conventional engineering sciences, compliance is an effect mostly tied to a negative behaviour. However, there is a continuous rise in applications which use compliance as the main source to enable a desired function. These compliant mechanisms are mostly used in micro manipulators because of their close to zero backlash and frictionless motion. Nevertheless it is possible to enhance the motion range of the compliant mechanisms by adjusting the size of the hinge. The subsequently introduced hinges show just an overview.

On the one hand there are hinges with lumped compliance. To allow flexibility in one or more directions the hinge blank is notched. The notch can have different geometries, for example a spherical, elliptical, parabolic or hyperbolic shape (Lobontiu et al. 2002) as well as a polynomial one (Zentner et al. 2011). A variation of the shape of the hinge not only affects the flexibility but also its stiffness.

On the other hand there are hinges with distributed compliance, for instance *leaf springs* (Zentner 2012). With their distributed compliance they have a larger twist angle than hinges with concentrated compliance (e.g. notch joints) (Dunning et al. 2011). To allow rotations with hinges based on leaf springs there are different possibilities as shown in Fig. 3. A combination of *leaf springs* (Trease et al. 2005), *free flex hinges* (Trease et al. 2005) also called *cross spring pivot* (Haringx 1949) and *cruciform hinges* (Trease et al. 2005).

The split-tube hinge also represents a joint with distributed compliance. In contrast to the previously mentioned hinges the movement is not realized by bending but by the torsion of a slotted cylinder (Fig. 4). This principle advantageously allows to design the axis of rotation in a very flexible way whereas the other axis maintain their stiffness. Thereby the parasitical movements of the joint are limited (Goldfarb and Speich 1999).

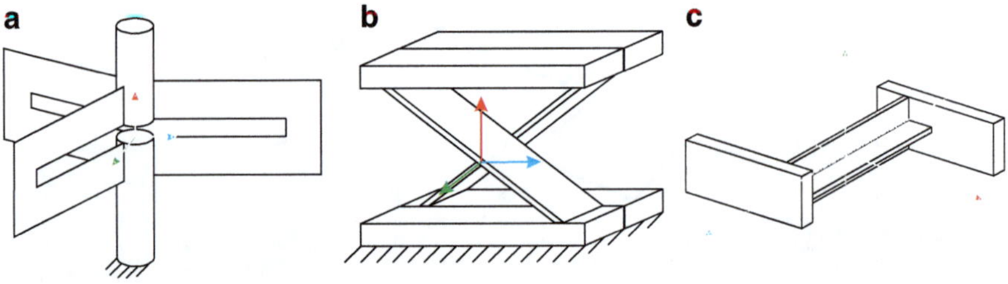

Fig. 3 (**a**) Leaf spring (Trease et al. 2005), (**b**) Free flex hinge (Trease et al. 2005) and (**c**) Cruciform Hinge (Trease et al. 2005)

Fig. 4 Split-tube hinge (Goldfarb and Speich 1999)

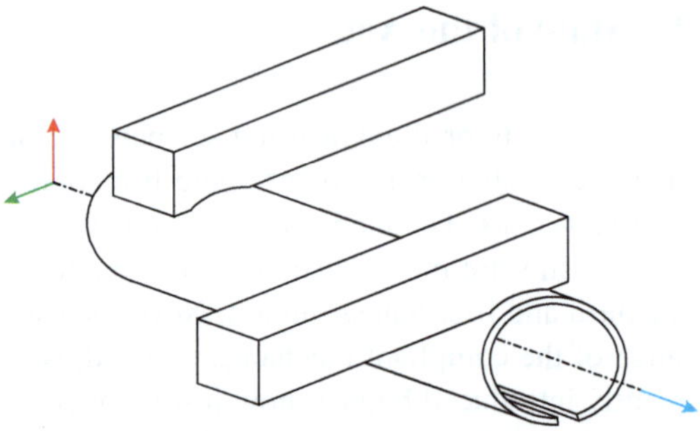

3 Requirements

For the development of a compliant robotic local structure, the requirements must be determined. The fundamental requirements can be derived from the already existing local structure (Fig. 2) which is to be replaced. Since the object motion is a spatial one with six degrees of freedoms (DOF) in total, the DOF of the local structure have to be the same. The three DOFs of rotation have to be independent from the hinge's structure. Furthermore injection molding and generative manufacturing methods are from further interest for such a monolithic joint. For this reason Acrylonitrile-Butadiene-Styrene (ABS) is used for the dimensioning. The material can also be used for three dimensional printing.

For the integration of the new structure into the surrounding of the PARAGRIP the structural and the dimensional synthesize have to be considered. Thus the measurements of the wrist joint are limited by a cylindrical area of Ø120 × 150 mm.

The kinematics and deformations of compliant joints not only depend on the desired displacement but even on the forces which have an effect on the mechanism. To enable the manipulability of the objects their weight has to be considered. Therefore the mass of the object is set to a maximum of $m = 1.5$ kg. The real loads depend on the grasp position of the wrist joint but in this paper only the maximal loads are considered.

The restoring torque is expected to keep the unloaded wrist joint in its initial position. For a robust design the restoring torque is set to T = 50 Nmm. This is two times the torque which can be realized by the existing two yoke springs.

Lifting the object is an important situation in the process of handling. At the beginning the object is always laying on the platform and the joints have to be deflectable in such a way that they are able to grasp it. For this the hinges have to provide an angle of at least $\alpha = 31.7°$ for the PARAGRIP where the current dimensions are taken into account. A requested requirement is to provide an even wider angle to enable further functions of the PARAGRIP such as infinitely spinning an object in the room. For a cube this means to have a possible angle of $\alpha_{wish} = \pm 45°$.

4 Structural Synthesis

Two main points are important for the presented compliant mechanism. First of all the hinge has to represent the state-of-the-art. Secondly, hinges from the state-of-the-art have to be combined to create the whole mechanism. Enlarging the possible motion by using multiple hinges of the same type is not taken into account. Hence, only one hinge for each rotational axis is used.

Spherical rotational joints, consisting of three orthogonal axes, can reach every point on a globe around the rotational center. With at least one oblique axis the movability can be less than a total globe. That is why the usage of oblique axes is inadvisable because of the limited area of motion of solid body hinges. They would reduce the area of motion and lead to an unsymmetrical one. Thus the two considered structures are the XYX-structure and the XYZ-structure (Fig. 5). They are named after the Euler-rotation.

If both robotic local structures are realized using the various hinges mentioned above the following mechanisms (Fig. 6) can be created: The introduced structures (h) and (i) as well as the second hinge of structure (b) are originated from Howell et al. (2013).

The different concepts are evaluated based on the criteria flux of force, compactness, area of motion and movement type. The flux of force rates the required connection elements between the particular hinges in the structure. The compactness takes qualitatively into account the dimensions of the entire structure. By the area of motion the possible twist angle is evaluated. The movement considers if it is a spatial or spherical structure and if the expected motion is symmetrical.

The evaluation leads to the three above-average concepts (Fig. 7) which are further dimensioned subsequently.

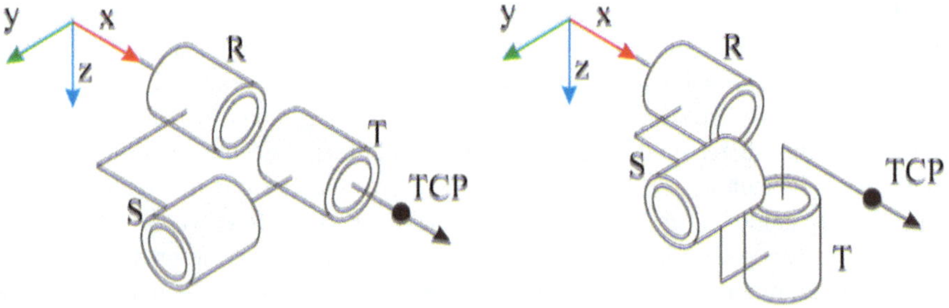

Fig. 5 Robotic local structure in XYX- and XYZ-structure

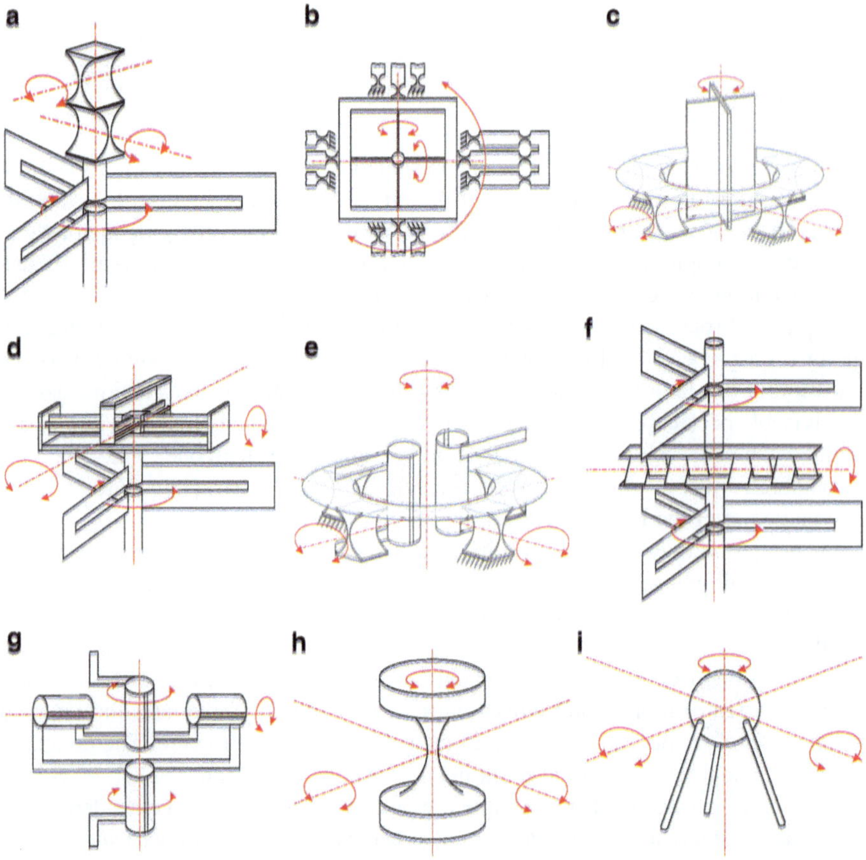

Fig. 6 Result of the structural synthesis

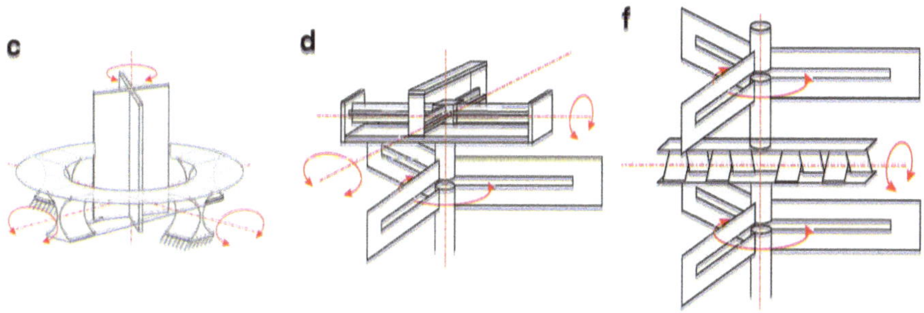

Fig. 7 Result of the qualitative evaluation

5 Dimensional Synthesis

The hinges used in the structural synthesis are considered separately because they are based on different concepts. The hinges are dimensioned in that way that the stress limit is not reached by the given restoring torques and deflections. At first the required resistance for the rotation of the hinge for the required deflection angle and the desired restoring torque is identified. That results into the product of the elasticity modulus E and the area moment of inertia I. The product EI is a function of the geometry and the material. Another boundary condition next to the required resistance is the acceptable tension. Since the area moment of inertia is dependent of the measurements of the cross section, possible combinations of the involved measurements are calculated leading to the required area moment of inertia I. Thereby, together with the weight and the twist angle, the effective stress (von-Mises) can be determined. The existing tension is calculated with a factor of safety ($j = 1.2$). The geometry with an effective stress beneath the critical one is selected. The procedure is representatively shown for a leaf spring. The calculation of area moment of inertia for the *free flex hinges*, the *cruciform hinges* and the *notch joints* can be taken from (Paros and Weisbord 1965; Trease et al. 2005; Haringx 1949).

The dimensioning of a leaf spring joint is based on a pseudo-rigid-body-model (PRBM) by (Howell et al. 2013; Howell 2001). For the schematic illustration of the deflected joint only one half of the leaf spring pair is highlighted (Fig. 8). On the left the first three models show the elements which connect the two springs. It is assumed to be rigid. Together with the boundaries the twist angle α leads to a inflection point P. The left side of the leaf spring is fixed by a rigid constraint to the connection element which is able to perform a rotation exclusively. The right side of the leaf spring fulfills a linear motion. For the usage of a PRBM the position of the inflection point P must be known. It is defined by the point where the bending changes its direction and therefore the torque within the cross section is zero. Independent from this exact position the PRBM can be derived. The measurements are determined later on.

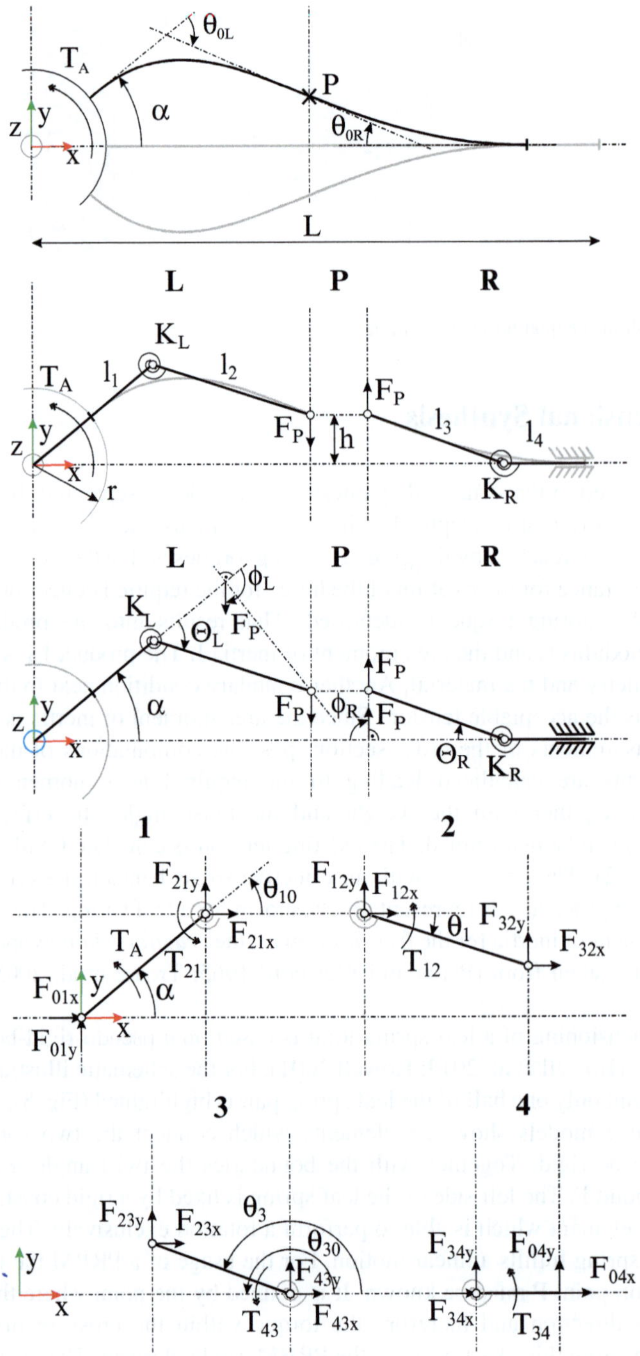

Fig. 8 Pseudo-rigid-body-model of the leaf spring

The PRBM can be divided in two areas left and right from the inflection point. Additionally to the axes and the moment of deflection the dimensions of l_1 up to l_4 can be described. The tensions of the original spring are represented by the spiral springs with the spring rates K_L and K_R. Only the inflection point force acts on the inflection point as there is no torque. Through the constraint on the right there is only a force in y-direction.

Using the PRBM by (Howell 2001) the forces and torques on both sides can be used to set up the balance of forces:

$$M_A = \gamma_L K_{\theta L} \frac{EI}{l_L} (\theta_1 - \alpha) \left(1 + \frac{(l_1 + r)\cos(\alpha)}{l_2 \cos(\theta_1)} \right) \tag{1}$$

To get the moment as a function of the deflection angle, the dimensions θ_1 and the lengths have to be determined. The angle θ_1 can be iteratively calculated by using two boundary conditions. One boundary condition is the height h of the inflection point P from both sides. The second one is the equilibrium of forces of the yoke springs at the inflection point. The lengths of the other sections can only be determined by a simulation due to the asymmetry of the PRBM. With the introduced equation, arbitrary values for the geometry and a given torque the angles of the PRBM can be evaluated. The lengths are adjusted in various steps until the inflection point of the PRBM and the inflection point of the simulation are equal. For the simulation the leaf spring with the same geometry, boundary conditions and torque is evaluated.

The comparison is realized by two equal lengths $l_R = 25$ mm and $l_L = 25$ mm. The result is a different inflection point $P_{25,25}$ in comparison to the simulation. Changing the lengths to $l_R = 28$ mm und $l_L = 32$ mm leads to the same result as in the simulation (Fig. 9).

All used hinges can be calculated in the same manner. Using this and the given restoring moment additionally the area moment of inertia can be evaluated depending on the angle.

Using the area moment of inertia the measurements can be calculated and a suitable combination can be selected (Fig. 10). By computing the tensions the measurements are checked afterwards. The calculation of the tension delivers the different dimensions and orientations of the joints in which no plastic deformation occurs.

With these figures the largest angle which does not conflict with the available space is selected. The measurements shown below refer to the directions shown in Fig. 11 (CAD-model). For the three calculated concepts Table 1 shows the dimensions.

The first concept is not usable due to the notch joints. The concentrated compliance provides a much smaller area of motion than the others. The third concept is also not usable for the PARAGRIP. Although its XYX-structure has no limitation in

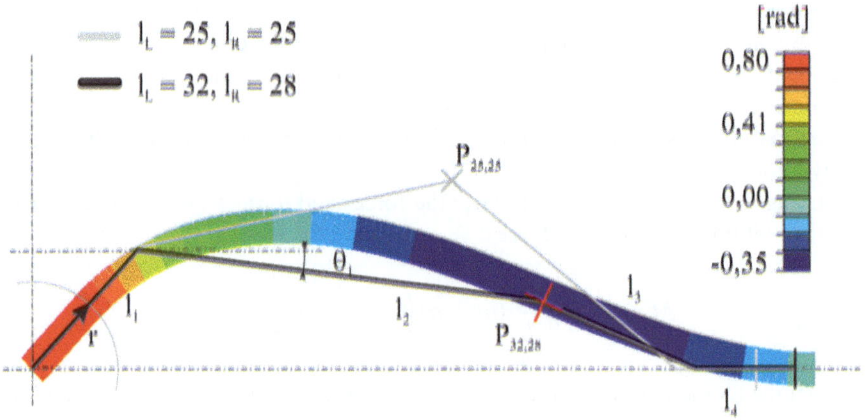

Fig. 9 Comparison of simulation and calculation

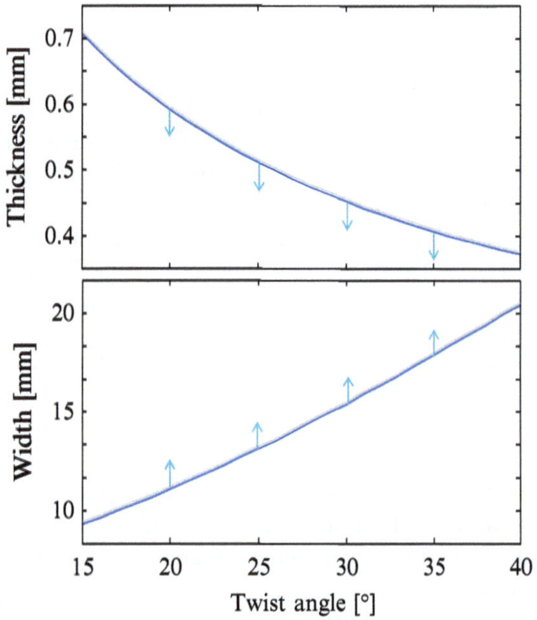

Fig. 10 Suitable combination of measurements for the area moment of inertia (Leaf springs)

general it is just reasonable in cases of wrist angles of 90°. Such angles do not appear in the presented joints so the XYX-structure is not applicable. The second concept fulfills all the requirements except the angel of 45°. Thus, this concept is selected for the further design.

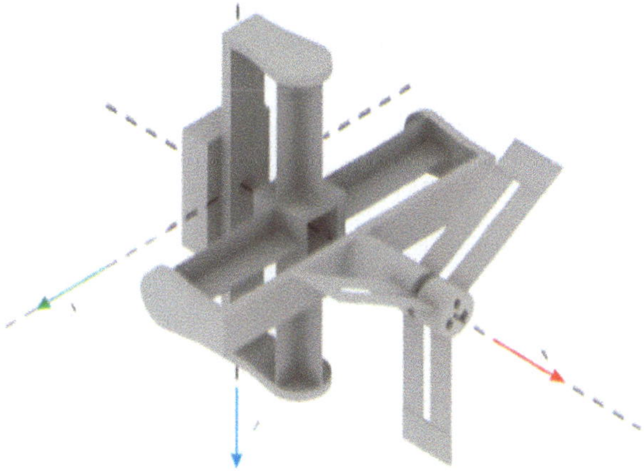

Fig. 11 CAD-model

Table 1 Results of the dimensional synthesis (dimensions of the selected structures)

		α [°]	x [mm]	y [mm]	z [mm]
(c)	Cruciform hinge	70	100	17.5	17.5
	Notch joint	18	60	50	67.7
	Notch joint	18	60	67.7	50
	Leaf spring	50	23.8	120	120
(d)	Cruciform hinge	40	26.7	120	26.7
	Cruciform hinge	40	26.7	26.7	102
	Leaf spring	50	23.8	120	120
(f)	Free flex	50	31.8	35	31.8
	Leaf spring	50	23.8	120	120

6 Design of the Compliant Robotic Local Structure

The designed CAD-Model is shown in Fig. 11. The connection elements between the hinges are designed more solid to restrict the rotation of the contemplated joints. In addition the two connections are illustrated. The connection to the regional structure at an angle of $x = 0°$ is shown on the left and three holes for the connection to the electromagnet are shown on the right.

7 Conclusions

The presented CAD-Model illustrates that the mostly in micro systems used hinges can also be used for macroscopic applications. To use a compliant wrist joint for the PARAGRIP, many restrictions have to be made. The calculations are only valid for the mentioned limitations of the objects weight. In comparison to conventional joints the compliant mechanism is restricted to a special area of application.

It is possible to use the joint in reality but existing motion tasks have to be reconsidered due to the limited angle of rotation. The angle of 45° cannot be reached. As a result it is not possible to infinitely turn the cube around by changing the end effector grasping points. A further optimization will finally enable this.

Acknowledgments As part of this work was carried out at the Cluster of Excellence at RWTH Aachen "Integrative production technology for high-wage countries (EXC 128)"; the authors would like to thank The German Research Foundation for their support.

References

Dunning AG, Tolou N, Herder JL (2011) Review article: inventory of platforms towards the design of a statically balanced, six degrees of freedom compliant precision stage. Mech Sci 2 (2):157–168. doi:10.5194/ms-2-157-2011

Goldfarb M, Speich JE (1999) A well-behaved revolute flexure joint for compliant mechanism design. J Mech Des 121(3):424. doi:10.1115/1.2829478

Haringx JA (1949) The cross-spring pivot as a constructional element. Appl Sci Res 1(1):313–332. doi:10.1007/BF02120338

Howell LL (2001) Compliant mechanisms. Wiley, New York. ISBN 9780471384786

Howell LL, Magleby SP, Olsen BM (2013) Handbook of compliant mechanisms. Wiley, Chichester. ISBN 978-1-119-95345-6

Lobontiu N, Paine JS, Garcia E, Goldfarb M (2002) Design of symmetric conic-section flexure hinges based on closed-form compliance equations. Mech Mach Theory 37(5):477–498. doi:10.1016/S0094-114X(02)00002-2

Müller R, Corves B, Hüsing M, Esser M, Riedel M, Vette M (2009) Rekonfigurierbares selbstoptimierendes Bauteilhandling. 8. Kolloquium Getriebe-technik Aachen 2009. Verlagshaus Mainz, Aachen, pp 297–311. ISBN 3-86130-984-X

Müller R, Riedel M, Vette M, Corves B, Esser M, Hüsing M (2010) Reconfigurable self-optimising handling system. In: Ratchev S (ed) IPAS 2010, IFIP AICT, vol 315. Springer, Heidelberg, pp 255–262. ISBN 978-3-642-11597-4

Paros JM, Weisbord L (1965) How to design flexure hinges. Mach Des 25:151–156

Riedel M, Nefzi M, Huesing M, Corves B (2008) An adjustable gripper as a reconfigurable robot with a parallel structure. In: Proceedings of the second international workshop on fundamental issues and future research directions for parallel mechanisms and manipulators, Montpellier, France, pp 253–260

Riedel M, Nefzi M, Corves B (2010) Performance analysis and dimensional synthesis of a six DOF reconfigurable parallel manipulator. In: IFToMM symposium on mechanism design for robotics 2012, Mexico City, 28–30 Sept 2010

Trease BP, Moon Y-M, Kota S (2005) Design of large-displacement compliant joints. J Mech Des 127(4):788. doi:10.1115/1.1900149

Zentner L (2012) Klassifikation nachgiebiger Mechanismen und Aktuatoren. In: Zentner L (Hrsg) Mechanismentechnik in Illmenau, Budapest und Niš, Technische Universität Ilmenau, Univ.-Verl, Ilmenau, 29–31 Aug 2012. ISBN 978-3-86360-034-1

Zentner L, Linß S, Erbe T (2011) On polynomial flexure hinges for increased deflection and an approach for simplified manufacturing. In: 13th IFToMM world congress in mechanism and machine science, Guanajuato, 19–25 June 2011

A Hexapod Walking Micro-robot with Artificial Muscles

I. Doroftei

Abstract As walking robots are requested to perform tasks in rough terrain, the development of actuators capable to flexibly adapt to the unstructured environment becomes more and more necessary. The conventional mechanisms with stiff joints make the robots more complex, heavy, large and expensive. Recent research in the field of materials discovered some new light and resistant alloys, which allow building compact, light and resistant articulated mechanisms. Such intelligent materials, called also artificial muscles, could be used to develop new actuators. Shape Memory Alloys are a category of artificial muscles which can be used as actuators for a walking robot. Even if these artificial muscles can exhibit large changes in shape when heated and cooled, only one part of their deformation can be used, if we want to maximize the actuator life. This is why smart mechanisms that can convert the small strain of the wire into large motion are necessary. In this paper, an example of using Shape Memory Alloys as actuators for a hexapod walking micro-robot is presented. Leg mechanisms that can convert the small strain of these actuators in large motion are also discussed.

Keywords Walking robot • Leg mechanisms • Shape Memory Alloys • Artificial muscles

1 Introduction

In order to miniaturize mobile robots and make them in a scalable manufacturing technique, mechanisms other than wheels are required for their locomotion systems. As we know, even with the most innovative suspension mechanisms, wheeled vehicles can negotiate obstacles with height at most twice their diameter. This is why the development of alternate locomotion mechanisms becomes quite necessary, especially in the case when the size of the mobile robot is reduced.

Even if wheeled platforms are more familiar in our daily life, legged vehicles are preferred in rough terrain because of their better mobility (Savant 2003). Highly

I. Doroftei (✉)
Mechanical Engineering Faculty, Theory of Mechanisms and Robotics Department, "Gheorghe Asachi" Technical University of Iasi, Iasi, Romania
e-mail: idorofte@mail.tuiasi.ro

E.-C. Lovasz et al. (eds.), *Microactuators and Micromechanisms*, Mechanisms and Machine Science 30, DOI 10.1007/978-3-319-15862-4_9

effective and power efficient biological legs are very common in nature. Legged animals in general and insects in particular can be found everywhere. They are well known not only for their speed and agility but also for their ability to traverse some of the most difficult terrains, climbing vertical surfaces, or even walking upside down (Kinglesley 2005).

As mobile robots are often requested to perform tasks in rough terrain, the development of actuators capable to flexibly adapt to the unstructured environment becomes more and more necessary. The conventional mechanisms with stiff joints make the robots more complex, heavy, large and expensive. In these conditions, the development of small and low cost actuators becomes desirable.

Different kind of actuators may be used for a robot driving. The so called conventional actuators (electrical, hydraulical and pneumatical actuators) are commonly used to drive actual robot systems. Recent research in the field of materials discovered some new light and resistant alloys, which allow building compact, light and resistant articulated mechanisms. Such intelligent materials, called also artificial muscles, could be used to develop new actuators (Doroftei and Adascalitei 2012; Doroftei et al. 2000; Torres-Jara et al. 2010).

A wide variety of artificial muscles have been investigated, Shape Memory Alloys (SMAs) being a category of such muscles, which can be used as actuators in the structure of a biomimetic walking robot. There are some properties of SMAs which have motivated their use as in a biomimetic actuation system: SMAs are compact, light-weight, with a high power to mass ratio; these actuators are direct-driven, so they are commonly used in the form of wires that are activated through resistive heating (by using an electric current) with no need for complex and bulky transmission systems; SMA actuators operate with no friction or vibration allowing extremely silent movements. They can exhibit large changes in shape when heated and cooled and can replace motors and solenoids for creating motion in many devices, even robots.

Shape memory alloys (SMA) based actuators have been successfully used in many robotic applications till now (Conrad and Mills 1997; Elahinia and Ashrafoun 2002; Esfahani and Elahinia 2007; Gilardi et al. 2010; Hoover et al. 2008; Huang et al. 2007; Kinglesley 2005; Price et al. 2007; Son et al. 2006).

However, these actuators also have limitations which need to be considered before applying them in any robotic system: a large portion of the power applied is dissipated as heat leading to low power efficiencies; the speed of actuation is dependent upon the rate of cooling of the wire leading to low bandwidths; even though SMAs exhibit relatively large strains (8 % for Nitinol), only a fraction of the net strain can be utilized in order to maximize the actuator lifetime; clever mechanism designs that convert the small strains into large motions are required; SMA contraction is highly non-linear owing to temperature hysteresis during joule heating of the material, this is why SMA actuators are difficult to effectively control (Mavroidis 2002).

In this paper, the design of actuators based on SMAs wires will be discussed and a mechanism that can convert the small strain of a SMA wire into large motion will be proposed. This mechanism will be compared with a serial one, in order to highlight its advantages. Also, the mechanism proposed here will be used as leg in the design of a small size hexapod walking robot. Numerical and virtual simulation of the compared legs and also, a virtual simulation of the proposed design of walking micro-robot using Ni-Ti SMA wires as actuators will be done.

2 Legs Based on SMAs Actuators

2.1 General Considerations

Generally, the number of degrees of freedom (d.o.f.) provided for each leg seems to be a compromise between an increased range of motion of the leg and a control simplicity. However, that turns out not to be the case. Leg designs having two d.o.f. or three d.o.f. have been considered for walking robot designs, since more than three d.o.f. provides no additional benefits comparing to the additional complications in control that would arise (Savant 2003).

Comparing with legs having three d.o.f., the legs with two d.o.f. provide a relatively simple control system, because there is one less actuator. But, at the same time, with only two d.o.f., it is very difficult to achieve the above-stated fixed goal of arbitrary navigation on a plane surface. These legs require a walking gait where the feet slip along the ground, if they are to travel in curved paths as well as travel straight ahead (Alexandre et al. 1998; Doroftei and Preumont 1999; Doroftei et al. 2011; Doroftei and Adascalitei 2012; Preumont et al. 1997). This is undesirable from a control standpoint, though, because systematic slippage of the feet implies that the robot's position no longer can be tracked by feedback from the motor controllers. Additional sensors would be necessary to track the motion of the robot's body over the ground. Although the trajectory of a robot using legs with two d.o.f. can never be a straight line because of the kinematics, the slippage does not cause any particular mechanical problem when the robot has a small weight (Alexandre et al. 1998; Doroftei and Preumont 1999; Doroftei et al. 2011; Doroftei and Adascalitei 2012; Preumont et al. 1997), which is the case of our design.

On contrary, with three d.o.f. per leg, the foot can be positioned anywhere within a three-dimensional workspace. This means that using six such legs the robot can navigate in either a straight or curved path over uneven terrain to reach any point on the terrain surface.

2.2 Actuators Design

In the case of an animal-like biomimetic walking robot design, actuators are the most critical components. Neuro-muscular systems of animals are using totally different control techniques comparing to those used to control conventional actuators in robotics. Animals ensure the leg joints motion by alternating the

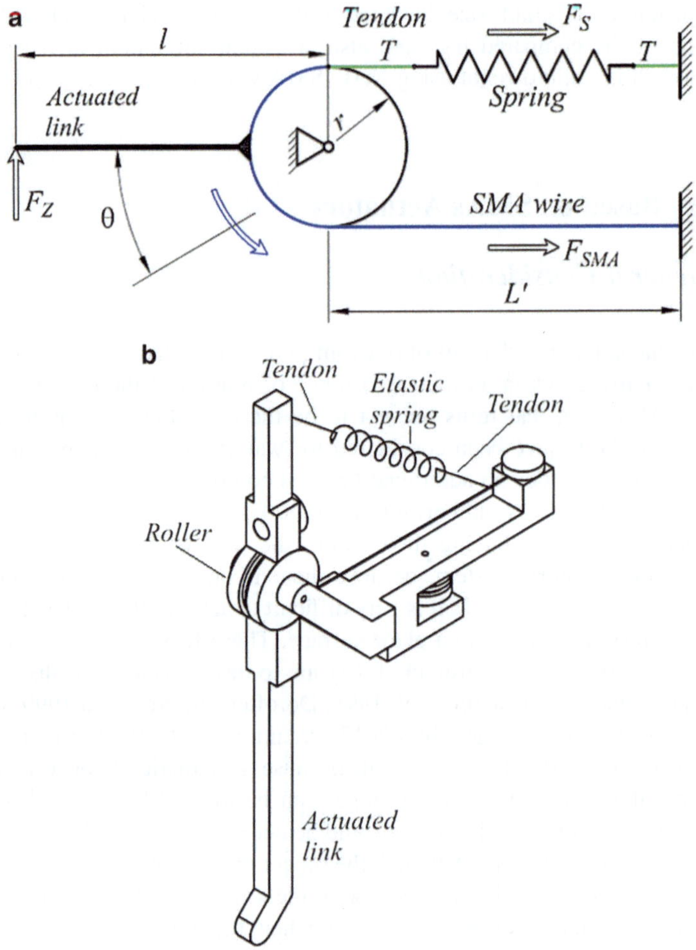

Fig. 1 Bias spring-type SMA based actuator: (**a**) kinematics; (**b**) possible constructive solution

contraction of an antagonistic muscles pair and regulate their stiffness by simultaneous contraction of these muscles (Alexandre et al. 1998). Because in the structure of biological walking mechanisms we find rotational joints, these kind of joints will be taken into account for our leg designs.

When designing a SMA based actuator, we have to know if the output force will produce one way or reversible rotation of the joint. Many SMA based actuator solutions may be proposed but only two of them have been used in the leg designs described in this paper. The first and the simplest one is the bias spring-type actuator (Fig. 1), using one SMA wire and an elastic spring (one way actuator), and the second one is a bidirectional actuator based on two SMA wires and two elastic springs (Fig. 2).

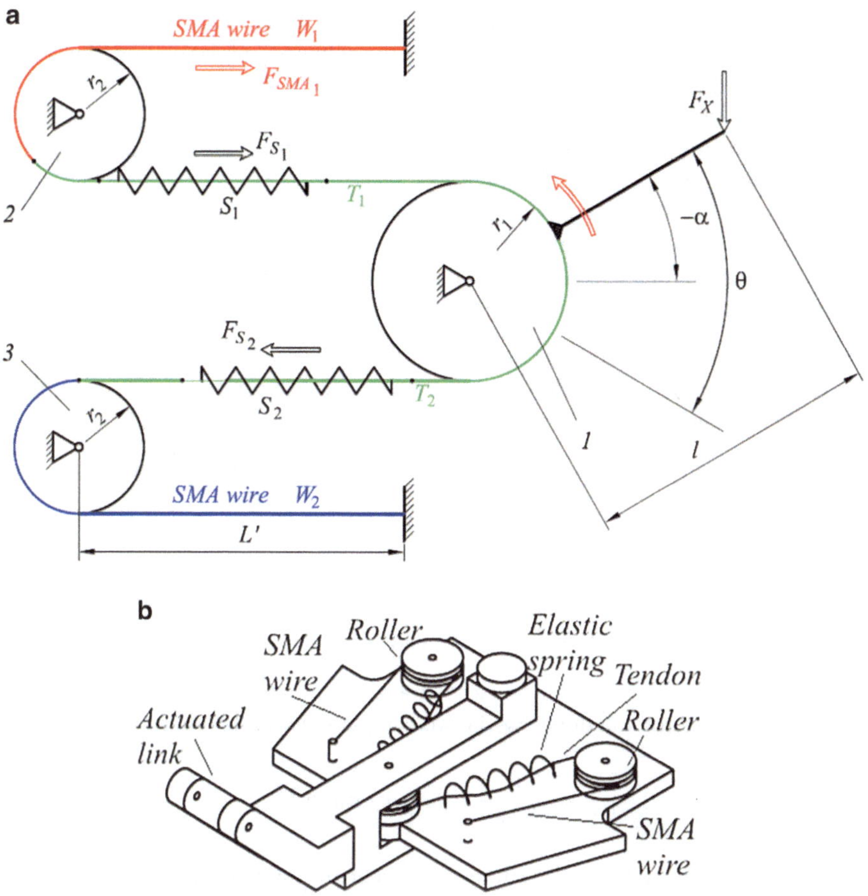

Fig. 2 Bidirectional SMA based actuator: (**a**) kinematics; (**b**) constructive solution

In the case of a hexapod walking micro-robot, a bias-spring actuator can be used to move the leg up and down (Fig. 1). Using it to rotate the leg around the vertical axis, the robot will only move in one direction (the force developed by the bias spring will not be capable to move the robot in opposite direction). In order to improve robot mobility and leg compliance, the second solution of actuator is used for this joint (Fig. 2), which is a bidirectional actuator based on two SMA wires, two elastic springs and two tendons, based on the solution adopted by (Gilardi et al. 2010). The spring S_2 biases the SMA wire W_1, whereas the spring S_1 biases SMA wire W_2. Counter-clockwise link rotation occurs by the contraction of W_1 wire whereas clockwise link rotation occurs by the contraction of W_2 wire. As the wire W_1 contracts, for example, the spring S_1 expands, absorbing the slack in the tendon T_1 until it is fully stretched and taut. Whereas the actuated link can rotate in counter-clockwise direction during the absorbing of the slack, depending of the spring stiffness, at the point where the tendon is stretched and taut, any further contraction of the SMA wire W_1 will act directly on the actuated link, rotating it in the mentioned direction. Simultaneously, as the actuated link is rotating, the spring S_2 expands and the slack in the tendon T_2 is absorbed. When the SMA wire W_1 is deactivated, the springs S_1 and S_2 will return to their original positions and the actuated link will return to its neutral position. When the wire W_2 is activated, the link is rotating in clockwise direction.

This solution, with spring-slack artificial tendons, mimics the nonlinear stiffness of the natural tendons. Whereas the nonlinear stiffness of the biological tendons limits the motion range of the natural links, the limit on the range of motion of the artificial joint is dependent on two factors:

- the SMA wire's stroke range must be sufficiently large to first absorb the slack in the active tendon to which it is connected in order to produce joint rotation;
- the link cannot rotate beyond the elastic limit of the opposing tendon.

Some defining criteria come into play when designing the SMA actuator:

- actuator stroke, which is done by the wire length;
- output force;
- wire diameter;
- number of wires;
- kinematics of the mechanism that converts strain of the wire in joint rotation.

If we impose the maximum actuator stroke, θ, the wire length can be computed. For the first actuator solution (Fig. 1), we can get the wire length as follow,

$$L = \frac{r \cdot \theta}{\varepsilon} = L' + r \cdot \pi \,[\mathrm{mm}], \tag{1}$$

where: $\varepsilon = 0.05 \div 0.08$ is the strain coefficient of the SMA wire; r is the roller radius, in [mm]; θ is maximum actuator stroke, in [rad]; L' is the distance between roller axis and the fixed end of the wire, in [mm].

It means, if a roller radius r is chosen, L' may be computed:

$$L' = \frac{r \cdot (\theta - \pi \cdot \varepsilon)}{\varepsilon} \,[\mathrm{mm}]. \tag{2}$$

For the second actuator solution (Fig. 2), if we consider that driven link (roller *1*, consequently) have a symmetric strokes according to neutral (horizontal) position of this link ($\alpha = \theta/2$), we get similar equations:

$$L_i = \frac{r_1 \cdot \theta}{\varepsilon} = L' + r_2 \cdot \pi \,[\mathrm{mm}] \tag{3}$$

and

$$L' = \frac{r_1 \cdot \theta}{\varepsilon} - r_2 \cdot \pi \text{ [mm]}, \tag{4}$$

where: $i = 1, 2$; r_1 is the radius of roller *1*, in [mm]; r_2 is the radius of roller *2*, in [mm].

The output force of bias-spring actuator may be computed as

$$F_{SMA} = \frac{F_z \cdot l}{r} \cdot \cos\theta + k(r \cdot \theta + \Delta x), \tag{5}$$

where: F_z is vertical contact force of the foot on the ground, k is the spring constant, Δx is spring elongation.

If we consider the counterclockwise rotation for a bidirectional actuator (see Fig. 2a), we can write

$$F_{SMA_1} = \frac{F_x \cdot l}{r_1} \cdot \cos\alpha + k(r_1 \cdot \theta + 2\Delta x), \tag{6}$$

where: F_x is horizontal contact force of the foot on the ground; α is counterclockwise angular stroke.

If the horizontal contact force, F_x, will be considered for the most unfavorable situation, that of the tripod gait ($F_x = \mu F_z$, with μ as friction coefficient between the foot and the ground) and if we consider that driven link have symmetric strokes according to neutral position of this link ($\alpha = \theta/2$), we get

$$F_{SMA_1} = F_{SMA_2} = \frac{\mu \cdot F_z \cdot l}{r_1} \cdot \cos\frac{\theta}{2} + k(r_1 \cdot \theta + 2\Delta x). \tag{7}$$

The motion generated by an actuator based on SMA wires is the result of the axial strain in the wire, due to the martensite/austenite transformation phase. Maximal axial strain, $\sigma_{\max}$, in a Ni-Ti wire is approximately 400 MPa. In these conditions, the maximum torques developed in a rotational joint, actuated by a number of n SMA wires with d diameter, is:

$$\tau_{\max} = n \cdot r \cdot \sigma_{\max} \cdot \frac{\pi \cdot d^2}{4}. \tag{8}$$

If a maximum torque, $\tau_{\max}$, is imposed, the number of wires with d diameter may be computed,

$$n = \frac{4\tau_{\max}}{\pi \cdot r \cdot d^2 \cdot \sigma_{\max}}. \tag{9}$$

or the wires diameter, d, results, if the number of wires is known,

$$d = \sqrt{\frac{4\tau_{\max}}{\pi \cdot r \cdot n \cdot \sigma_{\max}}}. \quad (10)$$

At first glance, if we analyze the last two equations, it would seem that the best would be to use a smaller number of larger diameter wires. However, the diameter of wire affects the reaction rate. This is why, to reduce the response time, a bigger number of smaller diameter wires is preferable to be used.

Assuming that the wire heat transfer is based on electrical heating inside and cooling by natural convection

$$m \cdot c_p \cdot \frac{dT}{dt} = \frac{U^2}{R} - h_C \cdot A_C \cdot (T - T_\infty), \quad (11)$$

where: R is wire electrical resistance; c_p is the specific heat; m is wire mass; ρ is wire specific density; A_C is area of the wire section; U is the electrical voltage applied to the wire extremities; T_∞ is the room temperature; T is the wire temperature; h_C is convection coefficient.

The response time of the wire may be computed as follow,

$$t = \frac{\rho \cdot c_p}{4h_C} \cdot d. \quad (12)$$

Equations (11) and (12) show that the use of larger wire diameter can generate a greater torque at the expense of a lower response time.

Knowing that the initial length of the wire can be determined using Eqs. (1) and (12) becomes as,

$$t = \frac{\rho \cdot c_p}{4h_C} \cdot \sqrt{\frac{4 \cdot \theta \cdot \tau_{\max}}{n \cdot L \cdot \pi \cdot \varepsilon \cdot \sigma_{\max}}} = K_w \cdot \sqrt{\frac{1}{n}}, \quad (13)$$

where K_w is a constant that depends on the physical properties of the SMA wire and the maximum torque required by the joint.

2.3 Leg Mechanism Design

Even though SMAs exhibit relatively large strains, only a fraction of the net strain can be used in order to maximize the actuator lifetime. This is why clever designs of mechanisms that convert the small strains into large motions are required, especially for moving the leg on the vertical plane. It means that we must pay special attention when we design the mechanism that convert the wire strain in displacement of the foot on vertical axis, in order to climb obstacles as high as possible, for a given wire length, or to use a wire length as small as possible, for a given height of the obstacle.

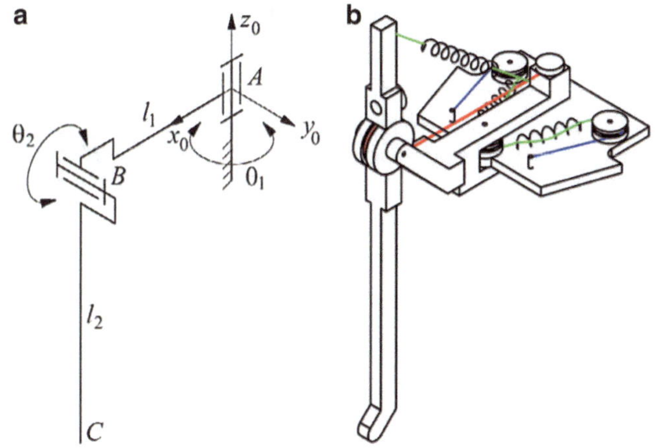

Fig. 3 Leg mechanism with two d.o.f.: (**a**) kinematics; (**b**) 3D CAD design

As mentioned in the previous paragraph, at the limit, the leg mechanism of a walking robot can have only two degrees of freedom. A leg solution with two d.o.f. (θ_1 and θ_2) could be based on a serial leg mechanism, using a bias spring-type SMA actuator and one bidirectional SMA actuator, as shown in Fig. 3. Because we want to amplify the small strains of the SMA wires, some pulleys which are fixed on each link of the leg are used.

When using serial mechanisms for the leg, the vertical displacement of the foot depends only on the roller radius and wire length. This is why, more complex and clever mechanisms should be found, in order to amplify this displacement when keeping the same roller radius and the same wire length. Such a mechanism is proposed in Fig. 4, (Doroftei and Adascalitei 2012; Doroftei and Stirbu 2014).

In order to prove the effectiveness of the second mechanism according to the serial mechanism, the equations that give us the vertical movement of the foot for both design solutions should be written. For the serial leg, this equation is as,

$$z_C = -l_2 \cdot \cos\theta_2, \tag{14}$$

while for the second mechanism, this equation becomes

$$z_F = -l_2 \cdot \cos\theta_2 - l_4 \cdot \cos\left[\tan^{-1}\left(\frac{a - l_2 \cdot \sin\theta_2}{b - l_2 \cdot \cos\theta_2}\right) + \tan^{-1}\left(\frac{a}{l_2 - b}\right)\right]. \tag{15}$$

Based on Eqs. (14) and (15), the curves of vertical stroke, Δz, for both mechanisms, have been represented in Fig. 5, for $\theta_2 \in [0° \ldots 30°]$. Next values of the constructive parameters have been considered in this numerical simulation:

- for the serial leg, $l_2 = 55$ mm;
- for tree structure leg, $l_2 = 16.75$ mm, $l_4 = 38.25$ mm, $l_2 + l_4 = 55$ mm, $a = 5$ mm, $b = 10$ mm.

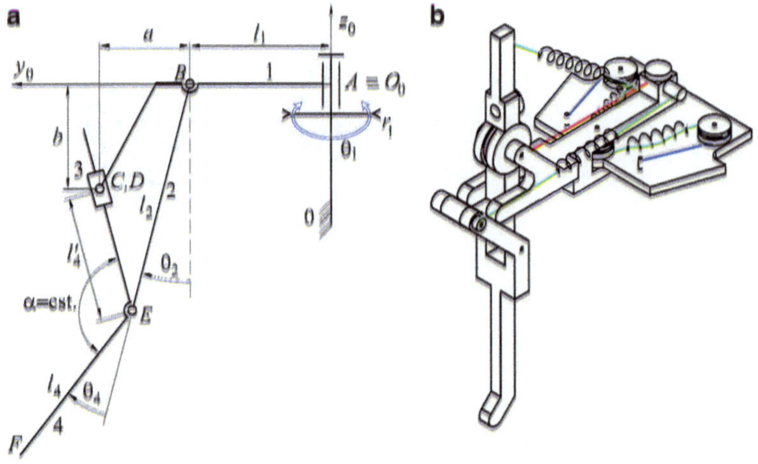

Fig. 4 Leg mechanism with tree structure: (**a**) kinematics; (**b**) 3D CAD design

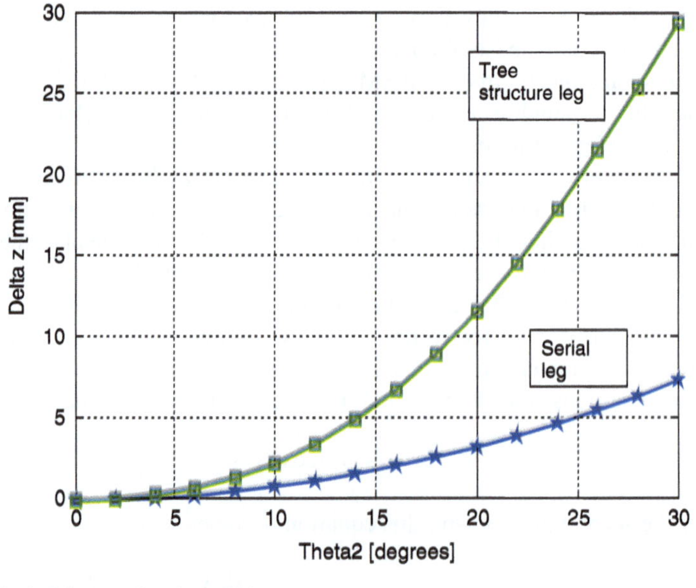

Fig. 5 Vertical foot stroke, $\Delta z = f(\theta_2)$

If we compare the two diagrams in Fig. 5, we can see that the leg with tree structure has a vertical stroke of the foot four times bigger than that of a serial leg structure when using the same length of the wire, which drives the θ_2 joint. In other words, if we want to get the same vertical foot stroke for both mechanisms, in the case of a leg with tree structure the length of the wire used for θ_2 joint should be four times smaller than that used for a leg with serial structure.

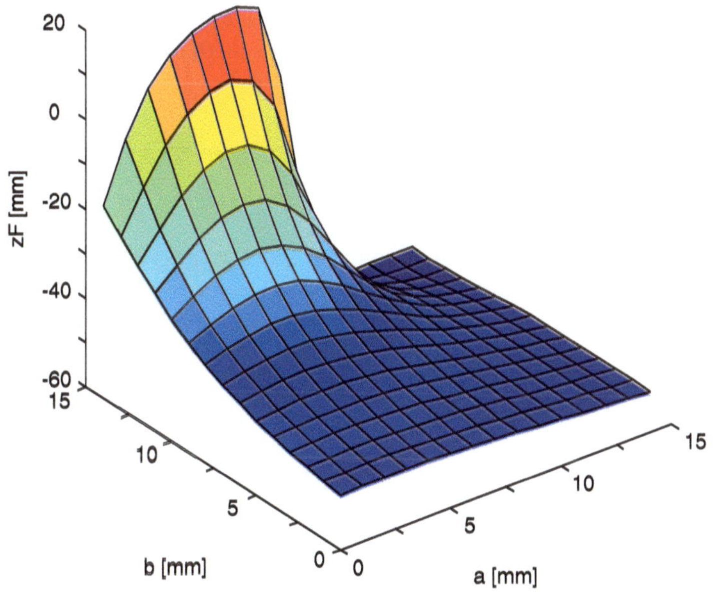

Fig. 6 The influence of a and b constructive parameters on the z_F coordinate

This will lead to less power consumption and smaller volume of the SMA actuator when using the second mechanism design. Or, if we consider using the same length of the SMA wire for both leg solutions, the robot using legs with tree structure could climb obstacles four times higher.

If the global dimensions of the leg based on the second mechanism are known (lengths l_1, l_2, l_4 are given), the vertical stroke of the foot is influenced by two other constructive parameters (a and b). The influence of these parameters on the coordinate z_F is shown in Fig. 6.

By doing the numerical simulation presented above, we have demonstrated the effectiveness of the second leg mechanism, with tree structure, vs. a serial leg mechanism.

The mechanism proposed here is only one of the solutions we may find for converting the small strain of a SMA wire into large motion. Finding these mechanism solutions is a matter of experience and ingenuity of the researcher.

Now, if we want to compute the torque for each joint of the mechanism shown in Fig. 4, in order to design SMA actuators, direct position kinematics of this leg will be written in Eq. (16). In these equations, the coordinates of the leg tip, F, according to O_0 origin of the referential frame are described.

$$\begin{cases} x_F = [l_1 + l_2 \sin\theta_2 + l_4 \sin(\theta_2 + \theta_4)] \sin\theta_1 \\ y_F = [l_1 + l_2 \sin\theta_2 + l_4 \cdot \sin(\theta_2 + \theta_4)] \cos\theta_1 \\ z_F = -l_2 \cos\theta_2 - l_4 \cos(\theta_2 + \theta_4) \end{cases} \tag{16}$$

where

$$\theta_4 = \tan^{-1}\left(\frac{a - l_2 \cdot \sin\theta_2}{b - l_2 \cdot \cos\theta_2}\right) + \tan^{-1}\left(\frac{a}{l_2 - b}\right) - \theta_2. \tag{17}$$

It means,

$$\begin{cases} x_F = \left\{ l_1 + l_2 \sin\theta_2 + l_4 \sin\begin{bmatrix} \tan^{-1}\left(\dfrac{a - l_2 \sin\theta_2}{b - l_2 \cos\theta_2}\right) + \\ + \tan^{-1}\left(\dfrac{a}{l_2 - b}\right) \end{bmatrix} \right\} \sin\theta_1 \\ y_F = \left\{ l_1 + l_2 \sin\theta_2 + l_4 \cdot \sin\begin{bmatrix} \tan^{-1}\left(\dfrac{a - l_2 \sin\theta_2}{b - l_2 \cos\theta_2}\right) + \\ + \tan^{-1}\left(\dfrac{a}{l_2 - b}\right) \end{bmatrix} \right\} \cos\theta_1 \\ z_F = -l_2 \cos\theta_2 - l_4 \cdot \cos\begin{bmatrix} \tan^{-1}\left(\dfrac{a - l_2 \sin\theta_2}{b - l_2 \cos\theta_2}\right) + \\ + \tan^{-1}\left(\dfrac{a}{l_2 - b}\right) \end{bmatrix} \end{cases} \tag{18}$$

If we derivate Eq. (16) by time, we get

$$\dot{\mathbf{X}} = \mathbf{J} \cdot \dot{\boldsymbol{\theta}}, \tag{19}$$

where:

$$\dot{\mathbf{X}} = [\dot{x}_F \quad \dot{y}_F \quad \dot{z}_F]^T, \tag{20}$$

$$\mathbf{J} = \begin{bmatrix} (l_1 + l_2 \sin\theta_2 + l_4 \sin\phi) \cos\theta_1 & (l_2 \cos\theta_2 + l_4 \cos\phi) \sin\theta_1 & l_4 \sin\theta_1 \cos\phi \\ -(l_1 + l_2 \sin\theta_2 + l_4 \sin\phi) \sin\theta_1 & (l_2 \cos\theta_2 + l_4 \cos\phi) \cos\theta_1 & l_4 \cos\theta_1 \cos\phi \\ 0 & l_2 \sin\theta_2 + l_4 \sin\phi & l_4 \sin\phi \end{bmatrix}, \tag{21}$$

$$\dot{\boldsymbol{\theta}} = [\dot{\theta}_1 \quad \dot{\theta}_2 \quad \dot{\theta}_4], \tag{22}$$

with $\phi = \tan^{-1}\left(\frac{a - l_2 \cdot \sin\theta_2}{b - l_2 \cdot \cos\theta_2}\right) + \tan^{-1}\left(\frac{a}{l_2 - b}\right)$.

In order to get the torque for each joint, next equation could be used

$$\boldsymbol{\tau} = \mathbf{J}^T \cdot \mathbf{F}, \tag{23}$$

where $\boldsymbol{\tau} = [\tau_1, \tau_2, \tau_4]^T$ is the torque vector and $\mathbf{F} = [F_x, F_y, F_z]^T$ is the vector of contact forces between the feet tip and the ground.

With τ_1 and τ_2 known, we may compute the wire diameter, for a known number of wires, or the number of wires, for a known wire diameter, using Eqs. (10) or (9).

3 Micro-robot Design

3.1 Kinematics

Based on the second leg mechanism discussed here, a solution of biomimetic walking micro-robot is proposed, using SMA as actuators and springs, in order to imitate the compliance of biological mechanisms. Kinematics of this robot is shown in Figs. 7 and 8. The conventional signs of the joint angles are represented in Fig. 9.

Direct kinematics of the robot expresses the coordinates of the feet tip with respect to the referential centre (geometrical centre of the robot body):

$$\begin{cases} x_{F_i} = l_{0_i} + [l_1 + l_2 \sin\theta_{2\_i} + l_4 \sin(\theta_{2\_i} + \theta_{4\_i})] \sin\theta_{1\_i} \\ y_{F_i} = l'_{0_i} + sign\left(l'_{0_i}\right) [l_1 + l_2 \sin\theta_{2\_i} + l_4 \sin(\theta_{2\_i} + \theta_{4\_i})] \cos\theta_{1\_i} \\ z_F = l''_0 - l_2 \cos\theta_{2\_i} - l_4 \cos(\theta_{2\_i} + \theta_{4\_i}) \end{cases}, \tag{24}$$

where: $i = 1 \ldots 6$ denotes the leg indices; l_{0_i} and l'_{0_i} are the distances between robot centre of mass and the axis of joints θ_{1_i}, on longitudinal and transversal direction; l_{0_i} is positive for $i = 1 \ldots 2$, $l_{0_i} = 0$ for $i = 3 \ldots 4$ and l_{0_i} is negative for $i = 5 \ldots 6$; l'_{0_i} is positive for $i = 1, 3, 5$ and l'_{0_i} is negative for $i = 2, 4, 6$; θ_{1_i} and θ_{2_i} are the angular strokes of the active joints; θ_{4_i} is computed using Eq. (17).

These coordinates are useful if we want to compute the contact forces between the feet tip and the ground. When we have these forces, we may get actuator force, using Eqs. (5, 6, and 7), or we may compute the actuators torques and, then, we may select the optimum SMA actuator (the diameter of the wire and the number of SMA wires).

3.2 Contact Forces

As a first step of the contact forces calculus, the force distribution for a given gait state of the robot can be computed, by expressing the static equilibrium equations of

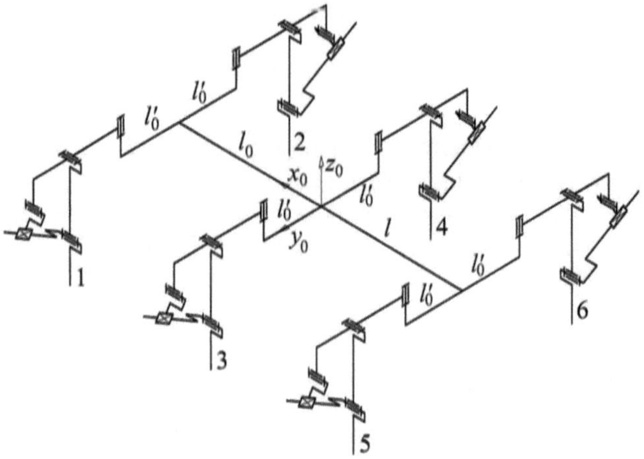

Fig. 7 Kinematics of the walking micro-robot based on legs with tree structure

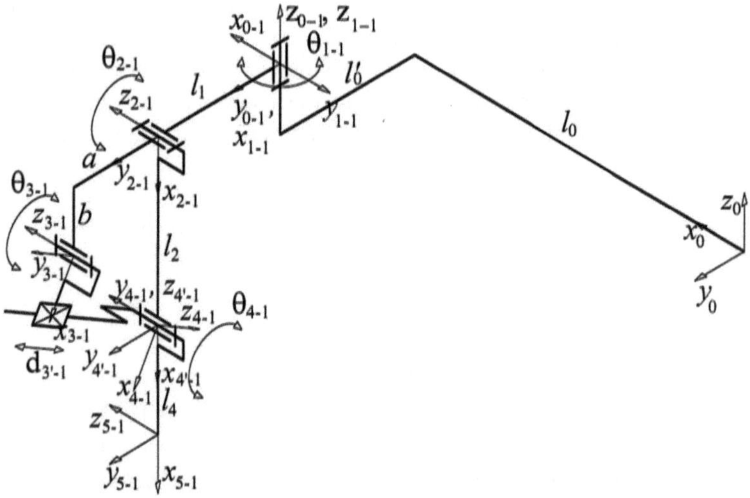

Fig. 8 Detailed kinematics of leg *1*

the whole vehicle. In order to simplify these equations, we assume that (Doroftei et al. 2011):

- the speed of the robot and its mass are small enough as well as dynamical effects could be neglected, which is, in fact, happening in our case;
- the ground support forces are vertical (Fig. 10);
- the mass of the leg link is negligible comparing to the mass of its actuators and also comparing to the mass of robot body. In other studies, the mass of all the leg is neglected. In our case, the mass of all the six legs is comparable with the mass

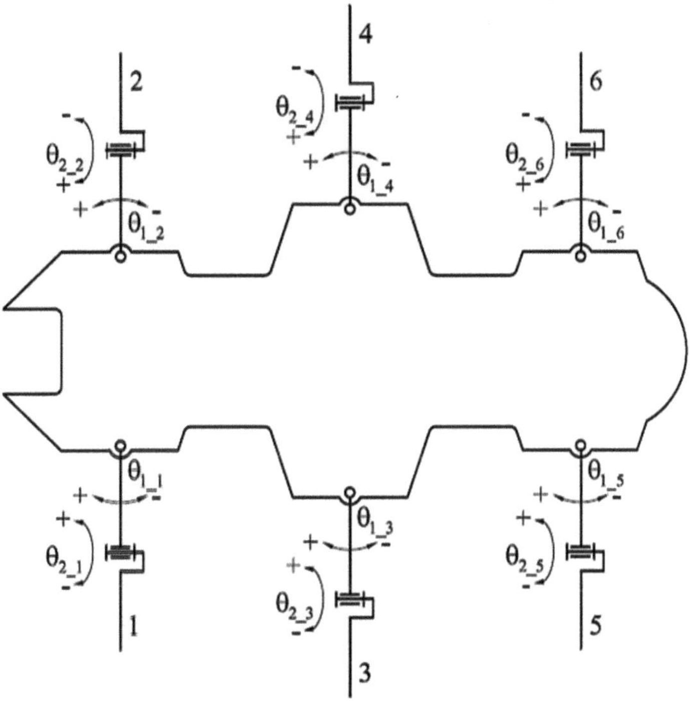

Fig. 9 Conventional signs of the joints angles, on a top view of the robot kinematics

of the robot body; this is why we consider that the influence of the leg mass should be taken into account.

Assuming that the mass of the robot body, M_b, its centre of mass (x_b, y_b), the mass of a leg, m_l, and the coordinates of their centre of mass $\left(x_{l_i}, y_{l_i}\right)$ are known, and assuming that all the six legs have the same mass, we may get:

$$\mathbf{A} \cdot \mathbf{F_z} = \mathbf{G}, \tag{25}$$

where:

$$\mathbf{A} = \begin{bmatrix} 1 & 1 & 1 & 1 & 1 & 1 \\ x_1 & x_2 & x_3 & x_4 & x_5 & x_6 \\ y_1 & y_2 & y_3 & y_4 & y_5 & y_6 \end{bmatrix}, \tag{26}$$

$$\mathbf{F_z} = \begin{bmatrix} F_{z_1} & F_{z_2} & F_{z_3} & F_{z_4} & F_{z_5} & F_{z_6} \end{bmatrix}^T, \tag{27}$$

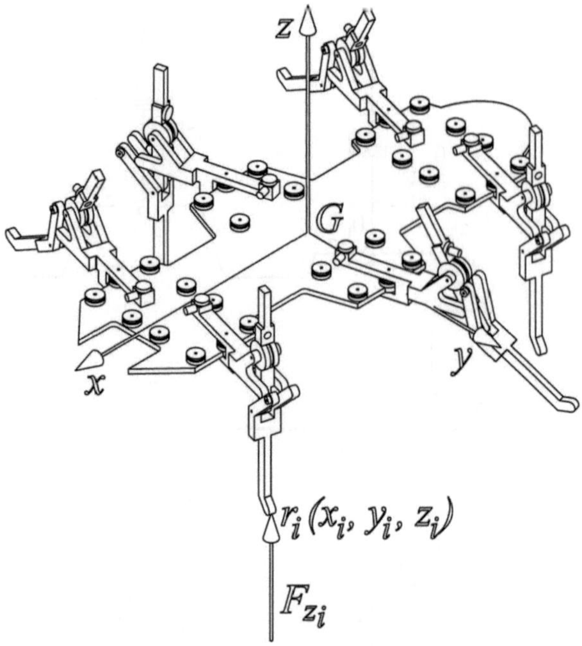

Fig. 10 Force distribution for a given gate

$$\mathbf{G} = \begin{bmatrix} (M_b + 6 \cdot m_l) \cdot g \\ m_l \cdot g \cdot \sum_{i=1}^{6} x_{l_i} \\ m_l \cdot g \cdot \sum_{i=1}^{6} y_{l_i} \end{bmatrix}. \tag{28}$$

Whenever there are more than three legs on the ground, this system of equations $(\mathbf{A} \cdot \mathbf{F_z} = \mathbf{G})$ is under determined (there are more unknown than equations); the minimum norm solution is given by

$$\mathbf{F_z} = \mathbf{A}^+ \cdot \mathbf{G}, \tag{29}$$

where $\mathbf{A}^+ = \mathbf{A}^T \left(\mathbf{A} \cdot \mathbf{A}^T\right)^{-1}$ is the pseudo-inverse of matrix $\mathbf{A}$.

3.3 CAD Model

Using the kinematics shown in Fig. 7, a constructive solution is proposed in Fig. 11, based on a leg mechanism with two d.o.f., actuated by one bias-spring SMA based actuator and one differential actuator. Thanks to the second actuator, the leg has an artificial compliance for horizontal movement.

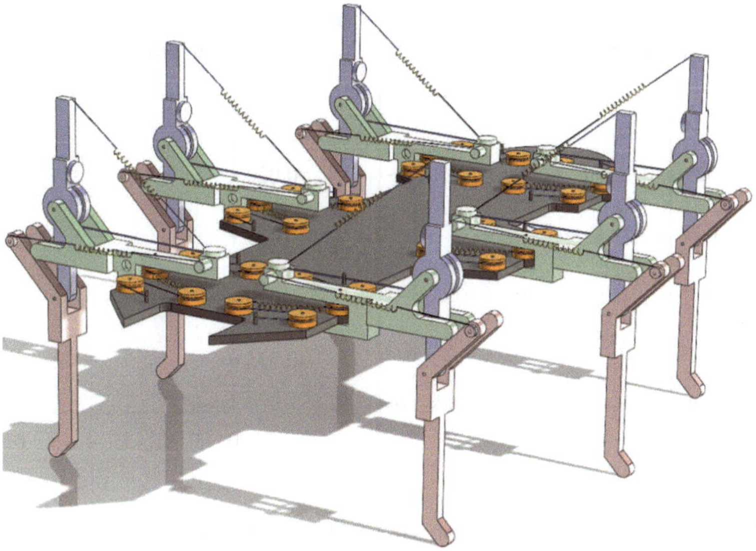

Fig. 11 Biomimetic walking micro-robot based on legs with tree structure mechanism

3.4 Virtual Simulation

A virtual simulation of the two solutions of leg has been performed in order to prove more the advantage of the solution adopted for our leg design. Also, a virtual simulation of the robot using the second leg solution, in order to have an idea about the contact forces between the feet and the ground. The same dimensional parameters, used for numerical simulation, will be used here.

Some of the results, the vertical movement of the leg tip, with respect to the referential centre (centre of *A* joint in Figs. 3a and 4a), for both leg mechanisms are presented in Fig. 12. These results confirm the numerical simulation, done above. For our virtual simulation of the entire robot, a 3D model of the solution shown in Fig. 11 has been used (see Fig. 13).

In the case of an alternating tripod gate (Fig. 14), the minimum number of legs in contact with the ground is three, being the most critical situation for SMA actuators. This is why, in order to get the contact forces, this kind of gait has been used in our simulation.

The alternating tripod gait is the best-known hexapod gait. A tripod consists of the front-back legs on one side and the middle leg on the opposite side. In this gait, the first and third leg on one side of the body move in unison with the contra-lateral middle leg, forming a moving tripod. This tripod alternates with the static, stable tripod made up of the remaining three legs. Since three legs are on the ground at all times, this gait is both "statically" and "dynamically" stable.

For θ_{1_i} and θ_{2_i} $(i = 1 \ldots 6)$ parameters varying according to Fig. 15, we have got the coordinates of the feet tip with respect to the origin of a referential attached

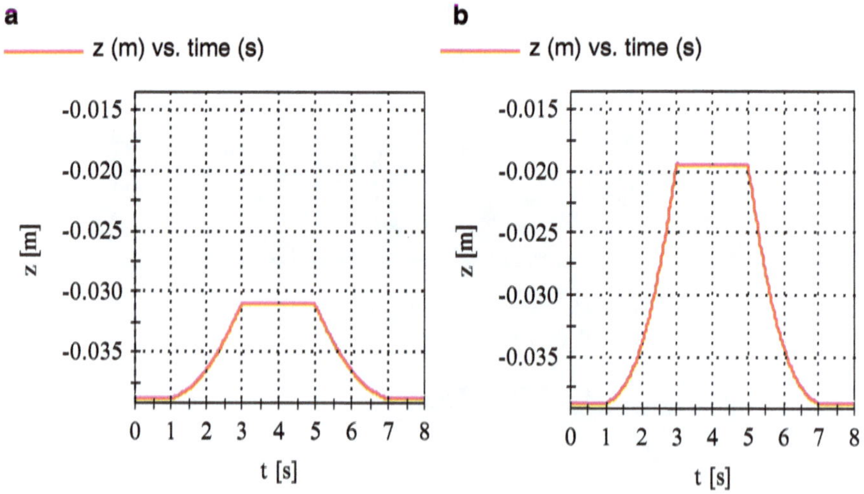

Fig. 12 Vertical movement of the leg tip: (**a**) serial leg; (**b**) tree structure leg

Fig. 13 Robot movement during virtual simulation

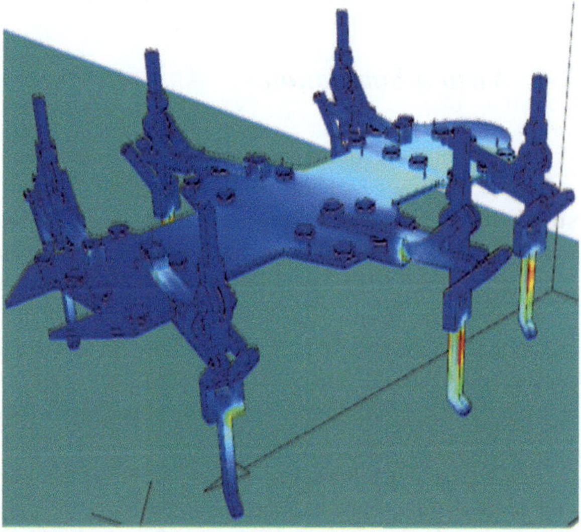

to the robot body. The convention of joints angles is shown in Fig. 9. We will further show these coordinates for the 6th leg only (see Fig. 16).

The 6th leg is in transfer phase during the first half of a robot step. This is why the diagram of the contact force between this leg and the ground looks like that presented in Fig. 17. The variation of this force during the second half of a step is due to the movement of the leg tip according to the robot body center. The contact force is bigger when the of the leg tip are closer to the robot body center (the coordinates x_6 and y_6 are smaller).

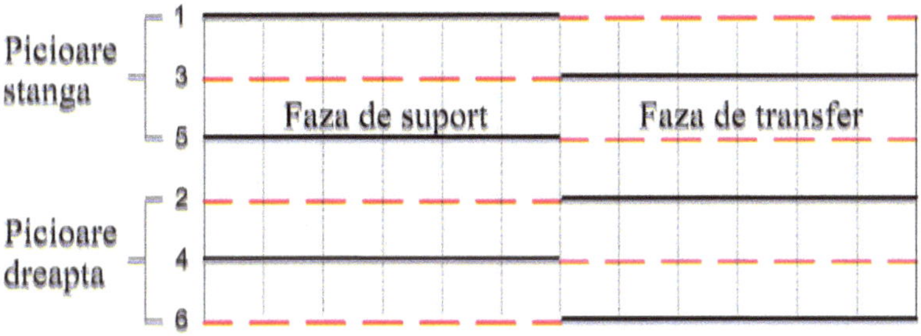

Fig. 14 Diagram of an alternating forward tripod gate

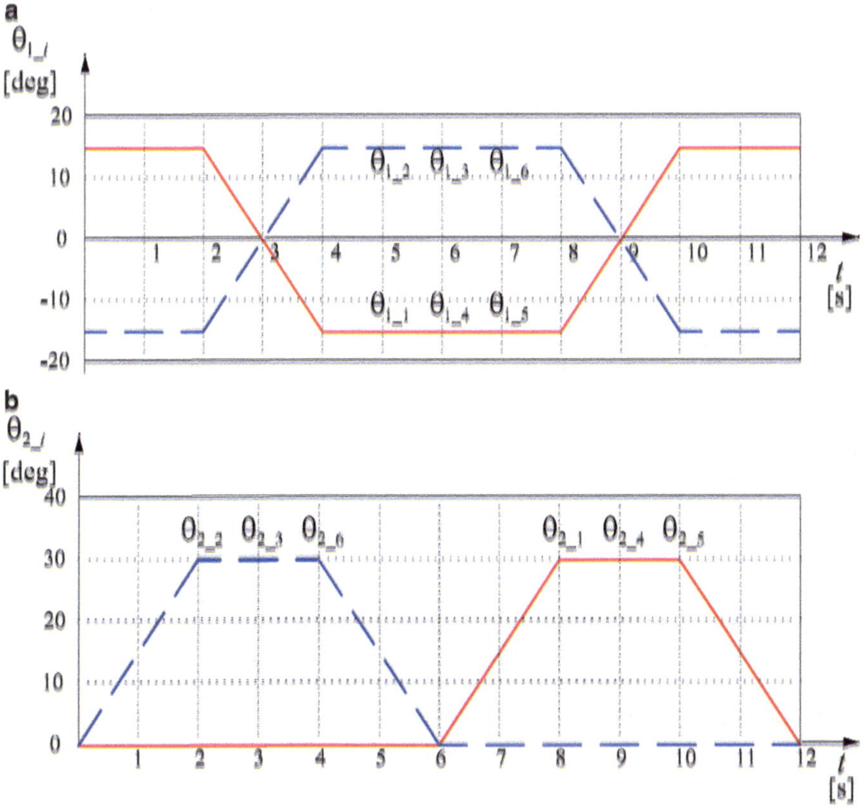

Fig. 15 Diagrams of θ_{1_i} (**a**) and θ_{2_i} (**b**) parameters for an alternating forward tripod gate

These simulation results may help the researcher to already have an idea about contact forces, before choosing the wire diameter, and also to compare with numerical results, avoiding some inherent mistakes.

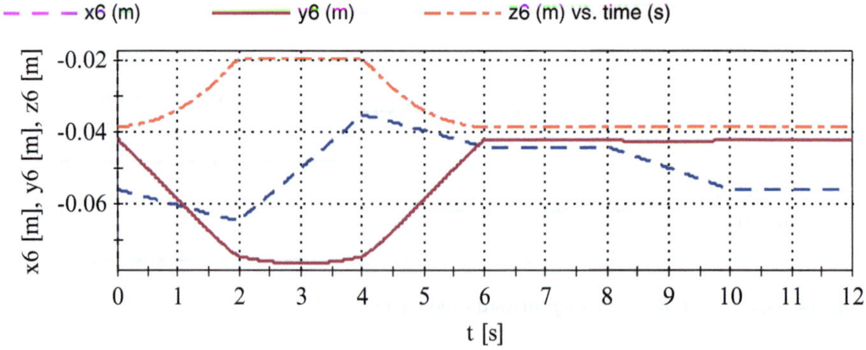

Fig. 16 Coordinates of the 6th leg tip with respect to the robot body center

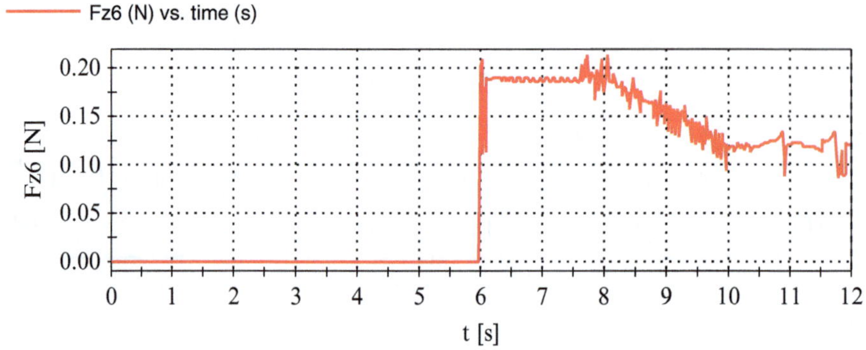

Fig. 17 Contact force between the 6th leg tip and the ground

3.5 Prototype

A picture of the real robot, using legs based on mechanisms with two d.o.f., discussed in this paper (see Fig. 4), is shown in Fig. 18.

The movement of the robot is achieved by heating small shape memory alloy wires (muscle wires) of 50 μm diameter, which are used for both actuators of one leg. Elastic rubber wires are used to return the links of the leg to the initial position, when the muscle wires are no more powered.

Although the robot trajectory can never be a straight line because of the simple kinematics, the slippage does not cause any particular mechanical problem because of the small weight of the vehicle.

All the legs are mounted directly on the electronic board without any other frames, in order to simplify the architecture of the robot. In order to increase the lengths of the wires (to increase the strokes of the legs) and to keep small overall dimensions for the robot, some pulley wheels are integrated in the structure of the leg.

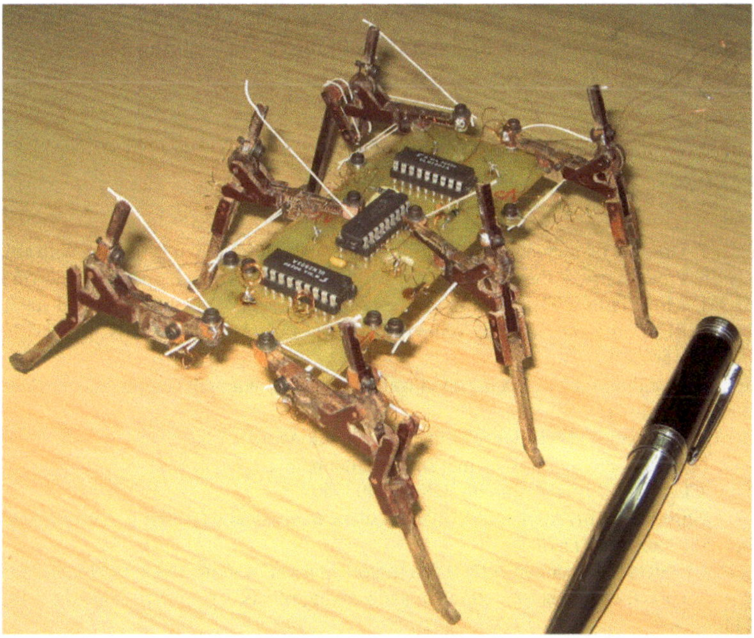

Fig. 18 Picture of the real robot with tree structure mechanism based legs

Thanks to a control board based on microcontroller, the robot can walk forward/backward and turn left/right, using an alternating tripod gait. Because not any encoder is used in the leg joints of this prototype, open-loop control tests have been performed to assess the effectiveness of the actuation mechanisms designed.

4 Conclusions

In order to miniaturize mobile robots and make them in a scalable manufacturing technique, mechanisms other than wheels are required for their locomotion systems. Thanks to their advantages in unstructured environment, legged vehicles would be more appropriate in this case. Recent research in the field of materials discovered some new light and resistant alloys, which allow building compact, light and resistant articulated mechanisms. Such intelligent materials, called also artificial muscles, could be used to develop new actuators, SMAs being a category of such muscles, which can be used as actuators in the structure of a biomimetic walking robot. Because of their limited strains, clever mechanism designs that convert the small strains into large motions are required. In this paper, the design of actuators based on SMAs wires has been discussed and a mechanism that can convert the small strain of a SMA wire into large motion was proposed. This mechanism was compared with a serial one, in order to highlight its advantages. Also, the mechanism proposed here has been used as leg in the design of a small size hexapod walking robot. Numerical and virtual simulation of the compared legs and also, a virtual simulation of the proposed design of walking micro-robot using Ni-Ti SMA wires as actuators have been done.

References

Alexandre P, Doroftei I, Preumont A (1998) An autonomous micro walking machine with articulated body. In: Proceedings of the 3rd IFAC symposium, IAV'88, Madrid, Spain, pp 557–562

Conrad JM, Mills JW (1997) Stiquito. Advanced experiments with a simple and inexpensive robot. IEEE Computer Society, Wiley-Blackwell, Los Alamitos. ISBN 10: 0818674083, 13: 9780818674082

Doroftei I, Adascalitei F (2012) A hexapod walking micro-robot with compliant legs. Appl Mech Mater 162:234–241

Doroftei I, Preumont A (1999) Development of an autonomous micro walking robot with articulated body. In: Proceedings of the 2nd international conference on climbing and walking robots, CLAWAR '99, Portsmouth, UK, pp 497–507

Doroftei I, Stirbu B (2014) Application of Ni-Ti shape memory alloy actuators in a walking micro-robot. Mechanics 20(1):70–79

Doroftei I, Cloquet JM, Preumont A (2000) A motorless micro walking robot. In: Proceedings of the 3rd international conference on climbing and walking robots, CLAWAR '2000, Madrid, Spain, pp 119–124

Doroftei I, Stirbu B, Plesu GH (2011) Force distribution for a walking robot with articulated body. In: Lovasz EC, Corves B (eds) Mechanisms, transmissions and applications, Mechanisms and machine science, Springer Netherlands, vol 3(2), pp 77–89

Elahinia M, Ashrafoun H (2002) Nonlinear control of a shape memory alloy actuated manipulator. Transanctions of the ASME. J Vib Acoust 124(4):566–575

Esfahani ET, Elahinia MH (2007) Stable walking pattern for an SMA-actuated biped. IEEE/ASME Trans Mechatron 12(5):534–541

Gilardi G, Haslam E, Bundhoo V, Park EJ (2010) A shape memory alloy based tendon-driven actuation system for biomimetic artificial fingers, part II: modelling and control. Robotica 28:675–687

Hoover AM, Steltz E, Fearing RS (2008) RoACH: an autonomous 2.4g crawling hexapod robot. In: Proceedings of IEEE/RSJ international conference on intelligent robots and systems, IROS 2008, Nice, France, pp 26–33

Huang HL, Park SH, Park JO, Yun CH (2007) Development of stem structure for flower robot using SMA actuators. In: Proceedings of IEEE international conference on robotics and biomimetics, ROBIO 2007, Sanya, China, pp 15–18

Kinglesley DA (2005) A cockroach inspired robot with artificial muscles. Thesis, Case Western Reserve University, Dep. of Mechanical and Aerospace Engineering

Mavroidis C (2002) Development of advanced actuators using shape memory alloys and electrorheological fluids. Res Nondestruct Eval 14:1–32

Preumont A, Alexandre P, Doroftei I, Goffin F (1997) A conceptual walking vehicle for planetary exploration. Mechatronics 7(3):287–296

Price AD, Jnifene A, Naguib HE (2007) Design and control of a SMA based dexterous robot hand. Smart Mater Struct 16:1401–1414

Savant A (2003) Leg design of walking robots. Fr. C. Rodrigues Institute of Technology, Mumbai University, Mechanical Engineering Department, Bs Report

Son HM, Gul JB, Park SH, Lee YJ, Nam TH (2006) Design of new quadruped robot with SMA actuators for dynamic walking. In: Proceedings of SICE-ICASE international joint conference, Busan, Korea, pp 344–348

Torres-Jara E, Gilpin K, Karges J, Wood RJ, Rus D (2010) Compliant modular shape memory alloy actuators composable flexible small actuators built from thin shape sheets. IEEE Robot Autom Mag 17(4):78–87

Side Chain Kinematics Simulation on Protein Conformational Changes

Mikel Diez, Victor Petuya, Mónica Urizar, Oscar Altuzarra, and Alfonso Hernández

Abstract Protein simulation remains as one of the most difficult task for biologists, physicists or engineers. The huge computational requirements of the protein models makes it difficult to obtain simulations of big conformational changes on protein structure. Side chain modelization is a critical step when obtaining the protein structure model. The time scale at which side chain movement occur can greatly increase the computational cost of the process. In this paper, we propose a side chain orientation procedure with a very low computational cost. This procedure has been implemented in combination with other procedures to obtain a computationally efficient simulation tool for protein simulation. This simulation tool has been used to simulate several protein conformational changes with success.

Keywords Protein • Side chains • Simulation

1 Introduction

As protein knowledge grows there is more evidence of the importance of conformational changes involving protein function. These conformational changes are usually part of the binding process of proteins with ligands or other proteins. Their understanding is essential to develop better and new medicines.

One of the most troublesome problems of protein simulation is the simulation of side chain movement. Side chains movement does not involve any conformational change on the structure of the protein, but may be relevant to allow certain conformational changes. The treatment of these side chains has been a widely discussed subject in the bibliography. In Levitt et al. (1996) it is proposed to treat side chain movement as a separate problem from main chain simulation. In Bohnenkamp et al. (2007) side chains size is reduced to limit its influence in the simulation. Later side chains are gradually resized to obtain the final structure. Other approaches propose to include only certain side chains degrees of freedom

M. Diez (✉) • V. Petuya • M. Urizar • O. Altuzarra • A. Hernández
Department of Mechanical Engineering, University of the Basque Country UPV/EHU, Bilbao, Spain
e-mail: mikel.diez@ehu.es

E.-C. Lovasz et al. (eds.), *Microactuators and Micromechanisms*, Mechanisms and Machine Science 30, DOI 10.1007/978-3-319-15862-4_10

into the simulation (Kavraki 2007). Finally in Levitt and Warshel (1975) side chains are modeled as spheres that contain each side chain atoms.

In this paper, we present a side chain orientation procedure with the objective of reducing the proteins potential energy. This procedure is combined with other simulation methodologies to obtain an efficient simulation package. The first one is a normalization procedure with the objective of homogenizing initial and final protein structures (Diez et al. 2011). The second one is a simulation procedure to obtain valid intermediate structures (Diez et al. 2014). And the final one, a third procedure that detects the presence of secondary structures in the protein and feeds the simulation algorithm with this information. The combination of these four processes allows to simulate protein motion with a low computational cost and always maintaining the kinematic continuity of the motion and the biological meaning of the obtained structures.

2 Kinematics of Proteins

To design a protein simulation program it is important to know the different time scales in which each possible movement inside a protein structure occurs. Taking this into account the following classification can be made:

- Local movements (10^{-15} to 10^{-10} s): In this time scale is where atom vibrations occur. Other movements that lay within this time scale are side chains internal movements.
- Rigid body movements (10^{-9} to 1 s): In this time scale lay secondary structure movements or protein structures whole sections movement.
- Macroscopic movements (10^{-7} to 10^{-4} s): Protein association-disassociation processes, secondary structure formation or folding processes happen in this time scale.

Attending to this classification, if proteins conformational changes shall be simulated, clearly only rigid body movements, and the degrees of freedom that produce them, must be taken into account. However, although side chain movements do not occur in the same time scale as rigid body movements, they are related to them. A proper side chain orientation may reduce the proteins overall potential energy, thus providing the simulation procedure with more energy for rigid body movement. Another important effect of side chain movement is that their reorientation may allow new pathways for protein conformational change.

To do the simulation, we use the kinematic model of the protein structure already used in our previous works (Diez et al. 2011) and represented in Fig. 1. In this model, only main chains dihedral angle rotations are considered as degrees of freedom of the proteins. The peptide bond angle, ω, is also considered to be constant, and its 0° or 180° value is ensured with the normalization procedure (Diez et al. 2011). Regarding secondary structures, these are going to be considered as rigid bodies throughout the simulation. To detect them, we use the procedure explained in Diez et al. (2013). In the next section, the procedure for side chain reorientation is explained.

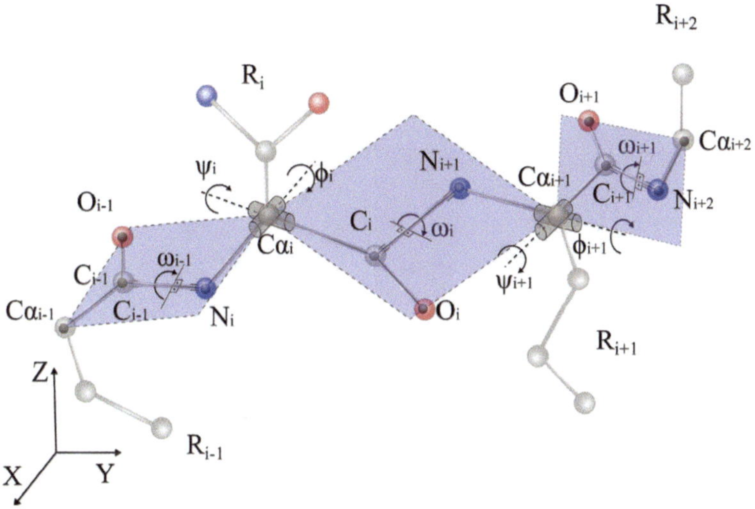

Fig. 1 Kinematic model of the protein main chain

3 Side Chain Orientation Procedure

The objectives to be fulfilled by the side chain orientation procedure are: First, reduce proteins potential energy value, and second, do it with a reduced computational cost. To achieve a low computational cost, only the rotational degree of freedom of $C_{\alpha i}$-$C_{\beta i}$ bonds will be considered (see Fig. 2).

One problem when attempting to reduce proteins potential energy is that it may lead to an energy minimization algorithm. This is unacceptable as it may produce an excessive increment on the computational cost. To avoid this problem, as it will be explained subsequently, the proposed procedure is based on the guidelines shown in Fig. 3.

The first step is to calculate the rotation direction of each side chain. To achieve this, each side chain is rotated in both possible directions calculating the produced energy increments E'_i and E''_i in the process. The optimal rotation direction is the one that has produced the greater reduction on the energy value. In case both rotation directions increase the energy value, the optimal one is the one that has produces less energy increment. After the optimal direction is determined, the side chain is returned to its initial stage. This determination of the rotation direction is done for each side chain.

Once every side chain rotation direction has been calculated the procedure attempts to reduce the protein potential energy. To reduce the computational cost of the procedure and not to develop an energy minimization procedure, the number of iterations has been limited to three. On each iteration, a different set of side chains are rotated. On the first iteration every, side chain is rotated 2°. On a second iteration again, every side chain is rotated but only 1°. On the final iteration only

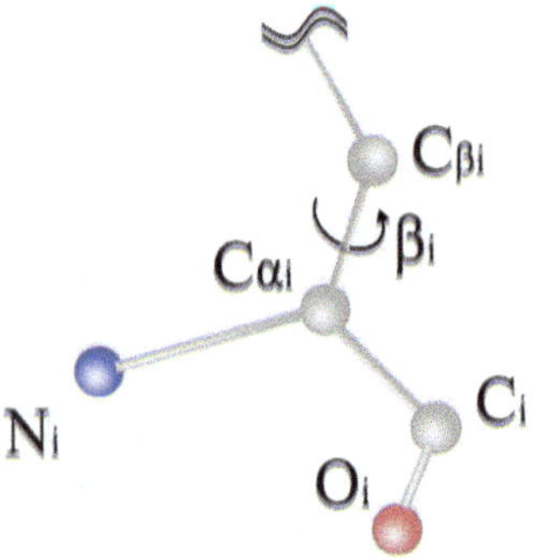

Fig. 2 Side chain degree of freedom

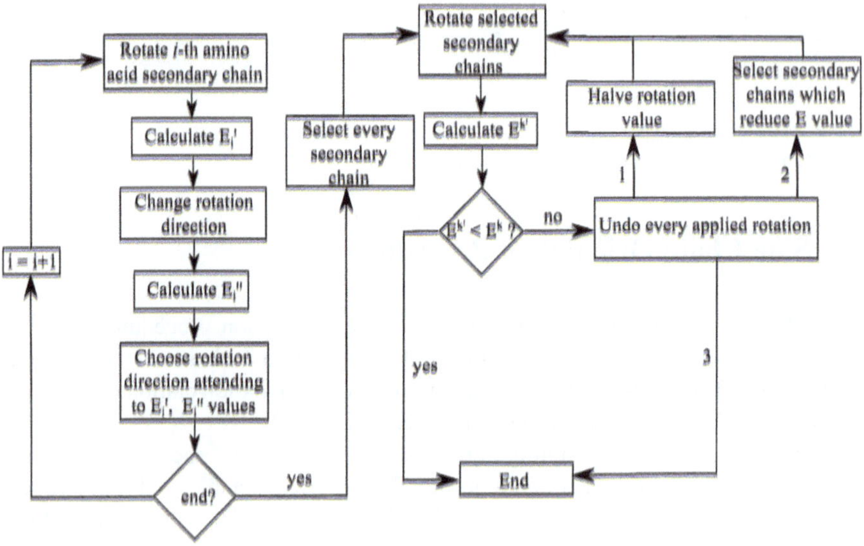

Fig. 3 Flux diagram of the side chain orientation procedure

side chains that have reduced the potential energy value when calculating their optimal rotation direction are rotated. After each iteration, a new value of the potential energy is obtained. These potential energy values are compared with the proteins potential energy before starting the side chain rotation procedure. If after any of the iterations the potential energy value is lower than the proteins initial one, the process ends, and the simulation continues. On the contrary, if after one iteration the energy value is greater than the initial one, applied rotations are undone and the next iteration starts.

It is important to understand that the rotation directions have been calculated independently for each side chain; thus, the effect of other side chains rotations has not been taken into account in that process. This has been done to maintain a reduced computational cost. This is also the reason of comparing the obtained potential energy after each iteration, especially on the final one. Although on the

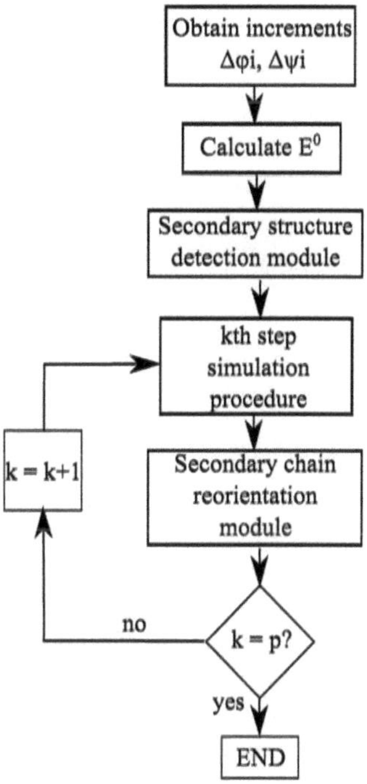

Fig. 4 Simulation process including the side chain orientation procedure and the secondary structure detection procedure

final iteration only the side chains that reduce the potential energy have been rotated, the effect of every rotation may not reduce the potential energy.

Once both procedures have been explained, it is essential to assess how to implement them on the simulation process. Side chain orientation procedure is mixed with the other processes; thus, it is important in which order they are executed during the simulation. To maintain the independence of the main simulation process from the side chain orientation process and secondary structure detection procedure, the next sequence of implementation is proposed according to Fig. 4.

As it is depicted in Fig. 4, side chain orientation procedure starts after the *kth* step simulation procedure has finished obtaining a valid structure. If the reduction of the potential energy of the protein is successful, there remains more energy for the next step of the simulation procedure.

4 Results

To assess the quality of the obtained structures, three indicators will be measured. Main chain atoms root mean square deviation (rmsd), Ramachandran plots (Ramakrishnan and Ramachandran 1965) and potential energy by means of the AMBER potential energy field (Cornell et al. 1995). These ensures the global

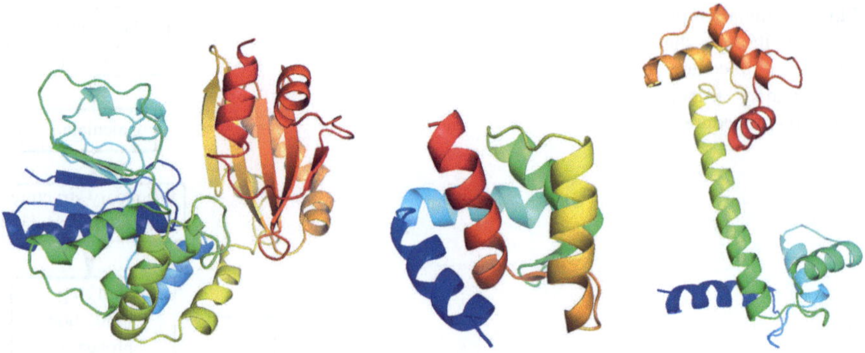

Fig. 5 *From left to right*: 1k20 protein, 1zac protein and 3cln protein

similarity between the structures (rmsd), the non-existence of collisions between atoms (potential energy) and the biological meaning of the obtained structures (Ramachandran plots). Intermediate data structures for rmsd comparison are obtained from the Morph server (Krebs and Gerstein 2000).

Three proteins are simulated. The first protein, type C inorganic Pyrophosphatase (family II) from Streptococcus gordonii protein, pdb (Protein Data Bank) entry 1k20 (see Fig. 5). The next protein is the Troponin C protein, pdb entry 1zac (see Fig. 5). Finally, the last protein is the Calmodulin protein, pdb entry 3cln (see Fig. 5). In order to test the effect of the different procedures included in the simulation process, three different strategies are proposed:

- Type 1: General simulation procedure and only the side chain orientation strategy will be used, not detecting secondary structures.
- Type 2: The three simulation processes are applied.
- Type 3: The same as Type 2 but with one exception: the side chains located on secondary structures will not be considered for the orientation procedure.

All simulations are run in a PC under Windows XP, with a Pentium core 2 duo 2.13 GHz making use of a single core.

In Table 1 results of the simulations of 1k20 protein are shown. The minimal rmsd value between the final data structure and the obtained final structure of 5.2 Å is obtained with Type 2 simulation. Biological meaning has been maintained throughout every simulation. This is confirmed by the Ramachandran plot values, always maintaining more than 90 % of amino acids inside the preferred zones (see Table 1). Finally, energy evolution during the simulation ensures the non-existence of collisions between atoms. With respect to the computational cost, results show how the side chain orientation module is the most expensive one. This can be perceived comparing type 1 and type 2 simulation strategies against type 3 simulation strategy. It can be observed how the exclusion of the side chains located on secondary structures reduces the simulation time.

In Fig. 6 rmsd and energy evolution can be seen. It can be appreciated how the energy value is reduced in the first steps of the simulation. This is possible thanks to

Table 1 1k20 protein results

Type of simulation	Energy %	rmsd Å	RP (% of atoms inside preferred zones)	Simulation duration (h)	Step duration (min)
Type 1	5.4	7.4	93	44	43
Type 2	4.4	5.97	93	37	43
Type 3	5.2	6.27	93	21	31

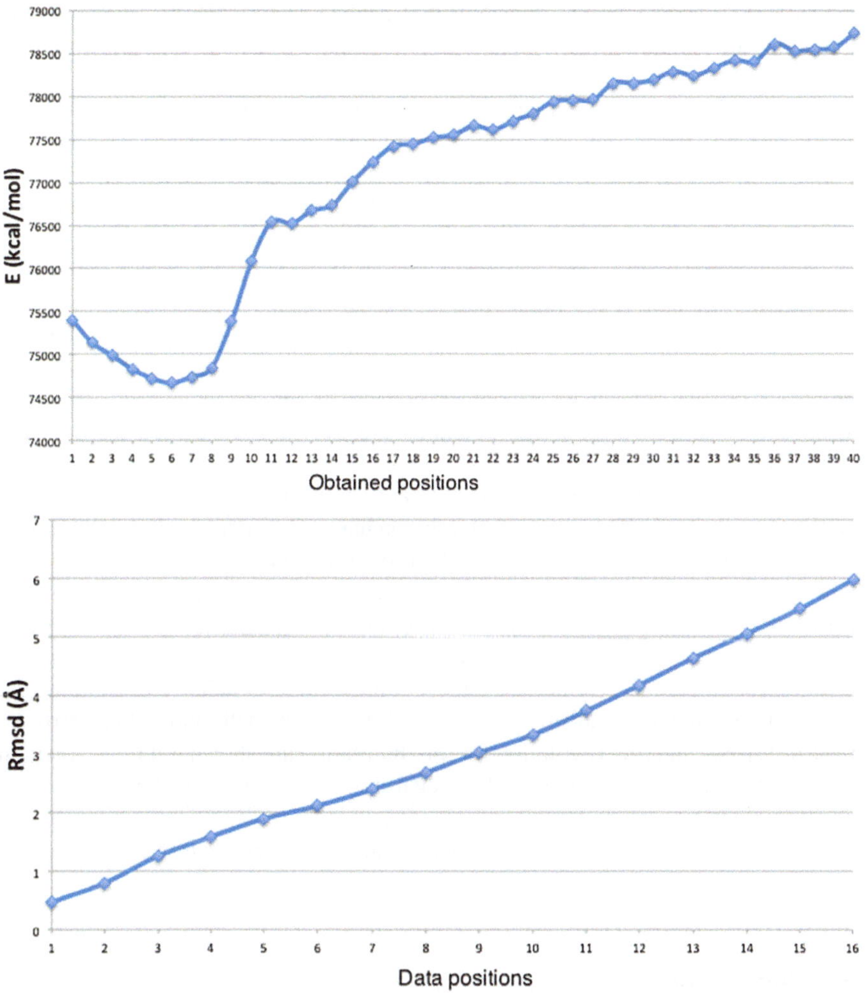

Fig. 6 Energy and rmsd evolution for 1k20 type 2 simulation

Table 2 1zac protein results

Type of simulation	Energy %	rmsd Å	RP (% of atoms inside preferred zones)	Simulation duration (min)	Step duration (s)
Type 1	3.8	3.15	98	64	121
Type 2	5.9	3.04	97	79	119
Type 3	4.4	2.52	96	87	97

the side chain reorientation procedure that achieves to reduce the energy value, providing the simulation with more energy for future steps and thus allowing new morphing paths.

On behalf of 1zac protein simulations, results are shown in Table 2. As with 1k20 protein there are no collisions during the simulation process and Ramachandran plot data indicates that intermediate structures maintain the biological meaning. In this protein minimal error has been obtained with type 3 simulation strategy with a final rmsd error of 2.52 Å. Finally, regarding the computational cost conclusions match the ones explained for 1k20 protein. Again the side chain reorientation module is computationally the most expensive procedure.

In Fig. 7 energy and rmsd evolution for 1zac type 3 simulation are shown. It can be observed how the energy value does not always rise. Again, this effect is produced by the side chain reorientation module presented in this paper that achieves to reduce the protein's potential energy value on certain steps of the simulation.

Lastly, 3cln protein simulation results are shown in Table 3. As can be seen neither type 2 simulation nor type 3 simulation has achieved to obtain a valid morph of the protein molecular mechanism. The problem resides on the secondary structure detection procedure. The molecular mechanism of this protein needs that the central α-helix has some degrees of freedom to allow the morphing. As those degrees of freedom are blocked in these simulations, the algorithm cannot obtain a valid protein path. Nevertheless, Type 1 simulation does achieve to simulate the protein molecular mechanism with a very low energy path and a final error of 6.34 Å. Energy and rmsd evolution for this simulation are shown in Fig. 8. Again in this case the most expensive procedure regarding the computational cost is the side chain reorientation process.

Finally, in Fig. 9, obtained final structures are compared against data structures.

5 Conclusions

Protein simulation always hits its limitations when trying to take into account every single degree of freedom that can be considered in a protein structure. Nowadays the computational cost related to this type of simulation is unmanageable for even actual supercomputers.

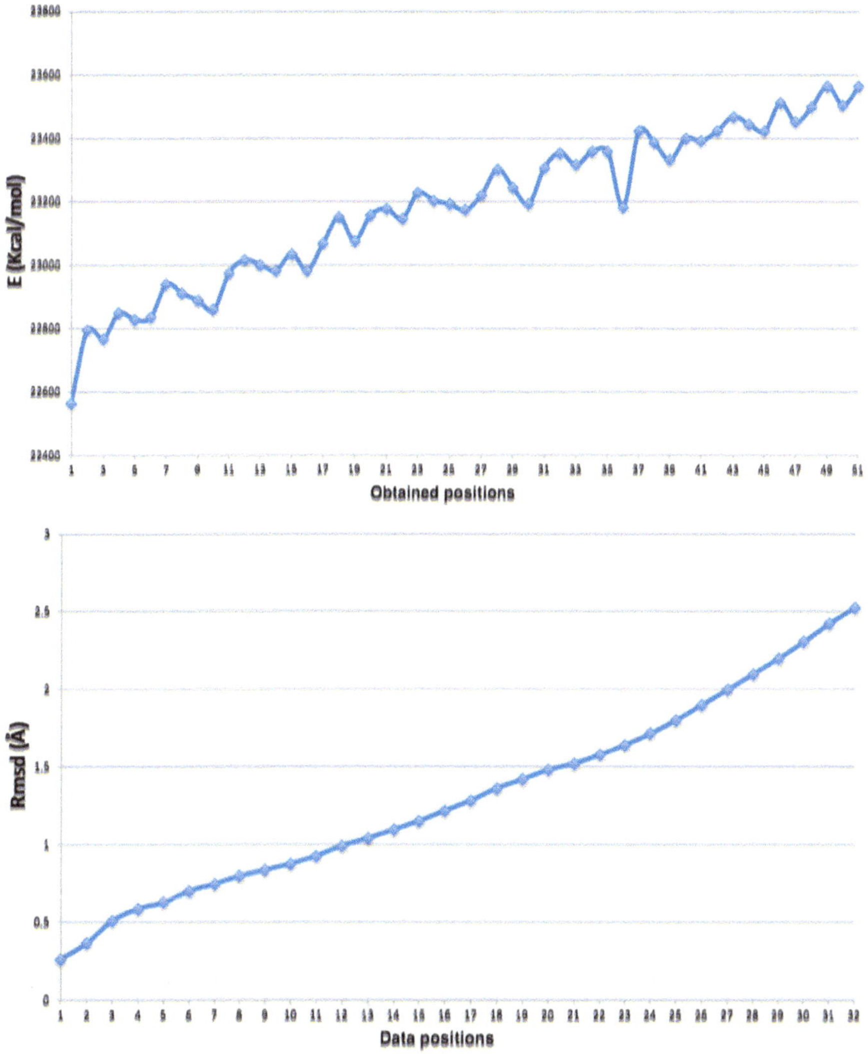

Fig. 7 Energy and rmsd evolution for 1zac type 3 simulation

Table 3 3cln protein results

Type of simulation	Energy %	rmsd Å	RP (% of atoms inside preferred zones)	Simulation duration	Step duration
Type 1	2.7	6.34	90	300 min	301 s
Type 2	–	–	–	–	–
Type 3	–	–	–	–	–

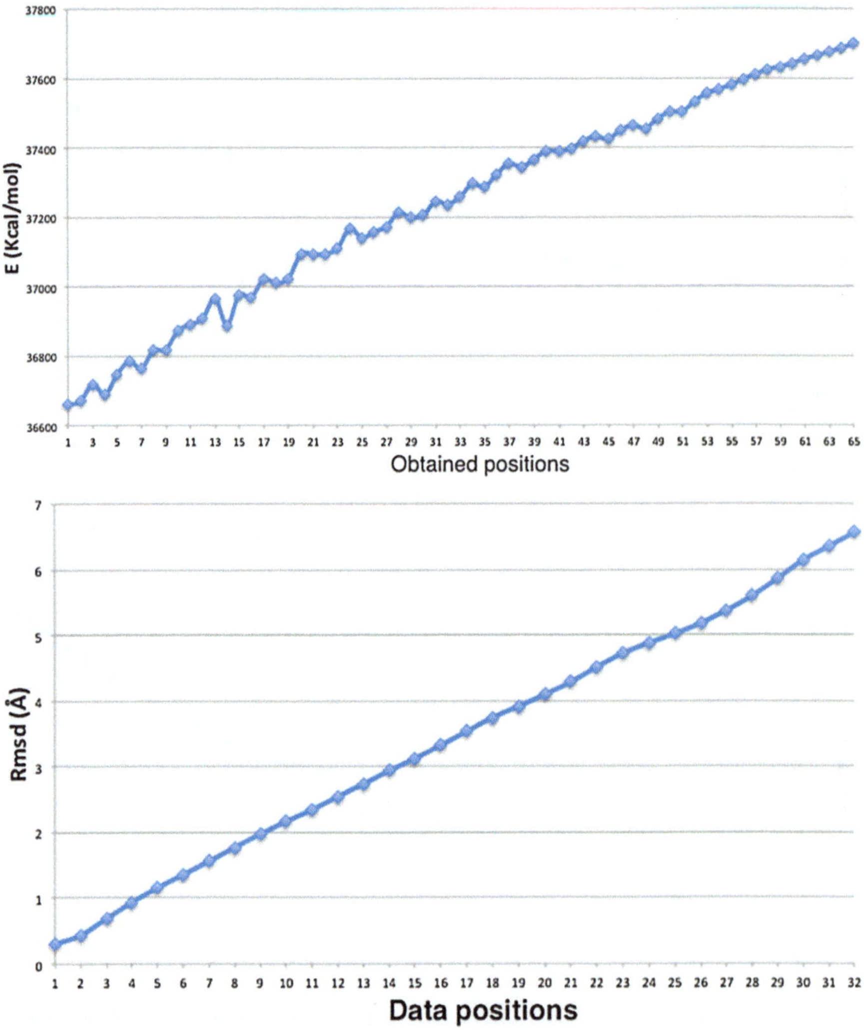

Fig. 8 Energy and rmsd evolution for 3cln type 1 simulation

In this paper we have presented a methodology to include side chains degrees of freedom into a protein simulation process. The methodology achieves to rotate side chains with a low computational cost, reducing proteins potential energy in the process. This procedure has been combined with other simulation and protein secondary structure detection and structure normalization strategies that together provide a reliable simulation tool with a reduced computational cost. Results show how the obtained protein morphs have low energy variations. As Ramachandran Plot data shows, obtained simulations also maintain the biological sense of intermediate structures. The procedure also ensures kinematic continuity of the obtained morph throughout the simulation.

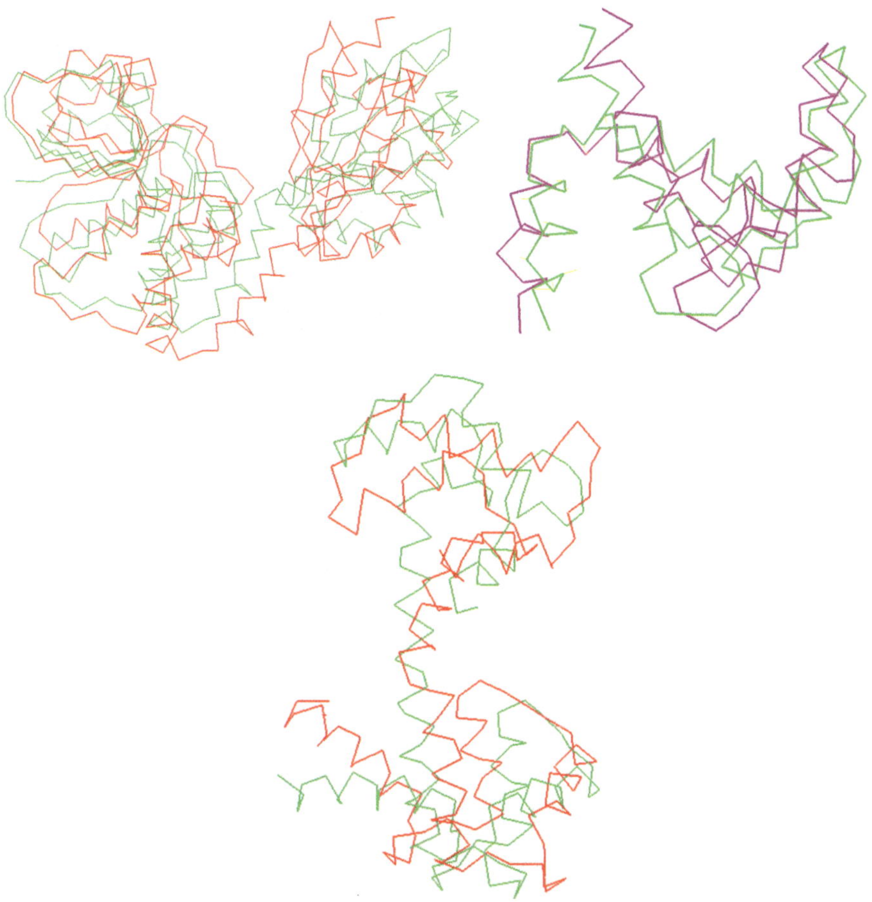

Fig. 9 Superposition of the obtained structures with data structures (*in green*)

Acknowledgments The authors wish to acknowledge the financial support received from the Spanish Government through the Ministerio de Economía y Competitividad (Project DPI2011-22955) and the Regional Government of the Basque Country through the Departamento de Educación, Universidades e Investigación (Project IT445-10) and UPV/EHU under program UFI 11/29.

References

Bohnenkamp P, Kazerounian K, Ilies HT (2007) Structural prediction of peptide based nano systems via progressive landscape evolution. In: Proceedings 12th IFToMM world congress, Besancon, France, pp 1–6

Cornell WD, Cieplak P, Bayly CI, Gould IR, Merz KM, Ferguson DM, Spellmeyer DC, Fox T, Caldwell JW, Kollman PA (1995) A second generation force field for the simulation of proteins, nucleic acids, and organic molecules. J Am Chem Soc 117(19):5179–5197

Diez M, Petuya V, Martínez-Cruz LA, Hernandez A (2011) A biokinematic approach for the computational simulation of proteins molecular mechanism. Mech Mach Theory 46 (12):1854–1868

Diez M, Petuya V, Martínez I, Hernandez A (2013) Protein secondary structure detection using dihedral angle parameters evaluation. In: Proceedings of SYROM 2013, Brasov, Romania, pp 127–134

Diez M, Petuya V, Martínez-Cruz LA, Hernández A (2014) Insights into mechanism kinematics for protein simulation. BMC Bioinforma 15:184

Kavraki LE (2007) Protein-ligand docking, including flexible receptor-flexible ligand docking. OpenStax CNX. Mar 22 2007. http://cnx.org/contents/4e7287b0-6c38-4829-abeb-3ae357bbf60f@10@10

Krebs WG, Gerstein M (2000) SURVEY AND SUMMARY: the morph server: a standardized system for analyzing and visualizing macromolecular motions in a database framework. Nucleic Acids Res 28(8):1665–1675

Levitt M, Warshel A (1975) Computer simulation of protein folding. Nature 253(5494):694–698

Levitt M, Gerstein M, Huang E, Subbiah S, Tsai J (1996) Protein folding: the endgame. Biochemistry 66:549–579

Ramakrishnan C, Ramachandran GN (1965) Stereochemical criteria for polypeptide and protein chain conformations. II. Allowed conformations for a pair of peptide units. Biophys J 5 (6):909–933

Design and Fabrication of Millimeter-Scale Crossed-Cylinder Wrist Mechanism with Two Degrees of Freedom

Brian D. Jensen, Jordan Tanner, Bryce Edmondson, Clayton Grames, Spencer P. Magleby, and Larry L. Howell

Abstract This paper describes the design and fabrication of a 2-DOF wrist mechanism suitable for fabrication with maximum dimension on the order of 2–4 mm. The design is based on the idea of 2 half-cylinders in contact such that their axes lie orthogonal to each other. In that way, each cylinder can roll parallel to the other cylinder's axis, giving 2 rotational degrees of freedom. To constrain the cylinders' motion, unique gear teeth are designed that allow rolling motion in either orthogonal direction, but constrain all other motions. Contact can be guaranteed using a compressive force acting to push the cylinders together. We first demonstrate the design at centimeter scale using FDM 3D printing. Based on the smooth motion achieved, we fabricate a wrist with maximum dimension of 3 mm using layered sheets of carbon nanotube composite material. Each sheet is individually patterned using photolithography.

Keywords Robotic wrist • Carbon nanotubes • Stacking assembly

1 Introduction

Wrist mechanisms are commonly used in a wide variety of grasping, cutting, and manipulating operations. They allow control of the angle of a tool with respect to a mounting shaft. Typically, the wrist mechanism is placed at the end of the shaft, immediately before the tool (such as a cutter or grasper) to improve the dexterity of the tool. Figure 1 shows a surgical tool produced by Intuitive Surgical, Inc., which uses a 2-degree-of-freedom (2-DOF) wrist to mount the tool.

Several examples of wrist mechanisms have been presented in the literature. The DragonFlex tool used two sequentially-mounted rotational joints to achieve 2-DOF motion (Jelinek et al. 2013). A similar wrist mechanism was described by Nai et al. (2011). A design based on planetary gearing, which decouples wrist and grasper motion, has also been demonstrated (Zhao and Nelson 2013). A design has

B.D. Jensen (✉) • J. Tanner • B. Edmondson • C. Grames • S.P. Magleby • L.L. Howell
Department of Mechanical Engineering, Brigham Young University, Provo, UT, USA
e-mail: bdjensen@byu.edu

E.-C. Lovasz et al. (eds.), *Microactuators and Micromechanisms*, Mechanisms and Machine Science 30, DOI 10.1007/978-3-319-15862-4_11

Fig. 1 A surgical tool from Intuitive Surgical, Inc. mounted on a 2-DOF wrist. The tool's shaft is approximately 8 mm in diameter

also been presented in small scale (2.5 mm diameter) using compliant mechanisms and lithographic patterning (Zoppi et al. 2008). However, the relatively complex design proved difficult to fabricate, and was only demonstrated in a 2X scale model.

Ideally, a wrist mechanism will allow large rotational motions (as large as $\pm 90^\circ$ is often desirable). In addition, the mechanism should operate with low swept volume to maximize maneuverability of the tool. Low friction and smooth motion are also important. Due to manufacturing constraints and the increased importance of friction at small scales, it has proven challenging to produce a wrist design with these characteristics appropriate for wrist size on the order of 2–4 mm.

This paper presents a design for a 2-DOF wrist mechanism capable of being manufactured with maximum dimensions on the order of 2–4 mm. The mechanism has minimal rubbing parts, resulting in low friction. We demonstrate fabrication at centimeter scale for testing using FDM 3D printing technology. We further demonstrate fabrication and assembly at the target size (3-mm diameter) using stacking of sheets of carbon nanotube composite material. Each sheet is lithographically patterned, and then assembled using a simple stacking procedure. The paper discusses further options for control of the mechanism.

2 Mechanism Design

The mechanism design is based on the concept of two cylinders in rolling contact with each other. A Compliant Rolling Contact Element (CORE) has previously been described (Halverson et al. 2010; Howell et al. 2013). The concept of the CORE is shown in Fig. 2. It uses two half cylinders (shown end-on in Fig. 2) with axes aligned parallel to each other. The half cylinders roll on each other to produce primarily rotational motion, though there is also some translation. One or more thin flexures are then placed between the half cylinders, as shown, to constrain them

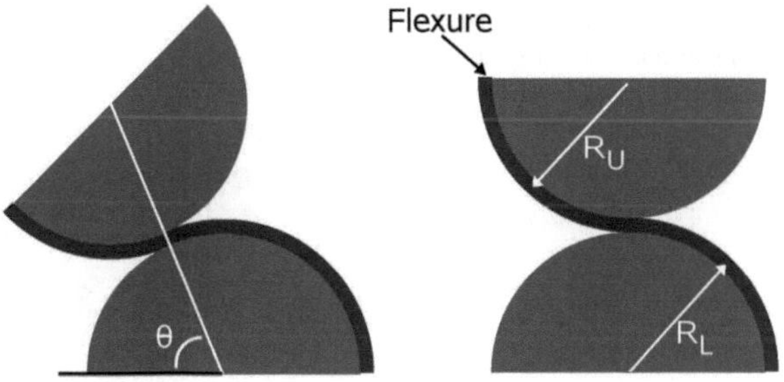

Fig. 2 Two positions of a CORE mechanism, consisting of two half cylinders (shown end-on in this drawing) that can roll on one another. R_U and R_L are the radii of the two cylinders

from any motion other than rolling. As a result, the CORE has one degree of freedom.

In contrast to the CORE concept, the crossed-cylinder wrist mechanism is achieved by rotating one of the cylinders so that the axes of each half cylinder are orthogonal to each other. In this configuration, a flexure would impede motion in one of the desired directions, so gear teeth are substituted to prevent torsion and shear. A compressive force between the half-cylinders then ensures that the two pieces stay in contact with each other. A rendered model of the full mechanism is shown in Fig. 3. As with the CORE mechanism, the two half cylinders experience only roll, and no sliding, with respect to each other. However, because the axes are orthogonal to each other, each half-cylinder rolls along a straight line on the surface of the other half cylinder, allowing 2 degrees of freedom (roll in either direction).

The gear teeth must be designed to function in two directions of roll. To accomplish this, each tooth has a cross-section of a rack tooth in the direction of the cylinder's axis, and the cross-section of a normal spur gear tooth in the orthogonal direction, as shown in Fig. 4. In this way, each gear tooth can fit into a gap on the mating cylinder. By staggering (offsetting) a row of teeth as shown in Fig. 5, the motion is constrained while allowing both cylinders to be identical, reducing the number of unique parts. By comparison, some previous 3D gear tooth designs have required different tooth design on each part (Hiller et al. 2011).

An important difference between the teeth in the crossed-cylinder mechanism and normal spur gear teeth is the load they are to carry. In normal spur gear teeth, the force always acts along the line of action, so that the most significant stresses in the tooth are often either shear stresses or bending stresses (Shigley and Mischke 1989). However, in the crossed-cylinder mechanism, the teeth must also support the compressive force acting to keep the two half cylinders in contact. Such stress could be analyzed using a finite element model.

Fig. 3 By rotating one half cylinder and adding gear teeth, the crossed-cylinder wrist mechanism is constrained to 2-DOF

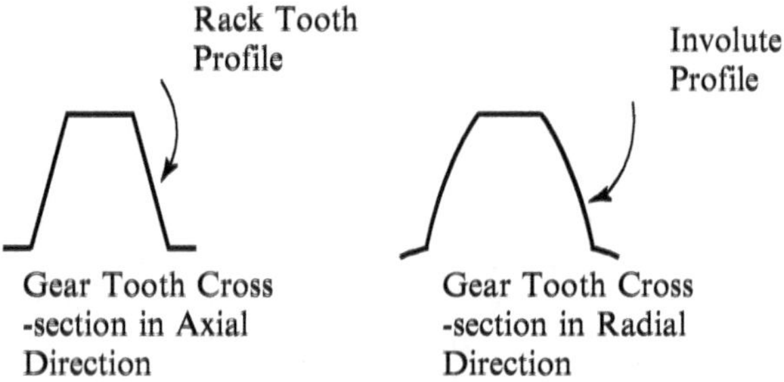

Fig. 4 The gear teeth are shaped like a rack tooth in the axial direction, but have an involute spur gear profile in the radial direction

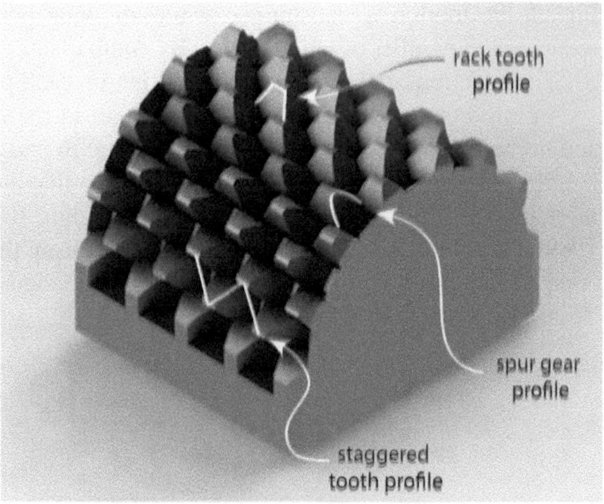

Fig. 5 The gear profile is staggered or offset in the axial direction to prevent motion due to torsion or shear

Fig. 6 Photographs of a prototype crossed-cylinder wrist fabricated using 3D printing [(**a**) – two geared half cylinders shown separately, (**b–d**) – two geared half cylinders in different positions]

3 Fabrication

To test the overall motion of the wrist, we fabricated a prototype at a relatively large scale using FDM 3D printing. The two geared half cylinders are shown separately in Fig. 6a. They are shown in three different positions in Fig. 6b–d, with an angle of the top surface of approximately 0°, 45°, and 90° with respect to the horizontal.

The mechanism shown is designed with a pitch diameter of 3 cm, and 18 teeth per full circumference (9 teeth per half-circumference). The designed pressure angle is 20°, and it has 3 % backlash. The addendum and dedendum were made using gear standards, so that the dedendum is slightly larger than the addendum. The mechanism was tested by placing the half cylinders together with their axes orthogonal to each other, as shown in Fig. 6b, and manually rotating the top half cylinder with respect to the bottom one. The motion was smooth without identifiable interference. The mechanism was able to rotate up to ±90° in both directions (along the bottom cylinder's axis, and around the bottom cylinder's circumference).

Based on the good performance achieved with the large-scale mechanism, we fabricated a wrist mechanism with maximum dimension (along the diagonal of the

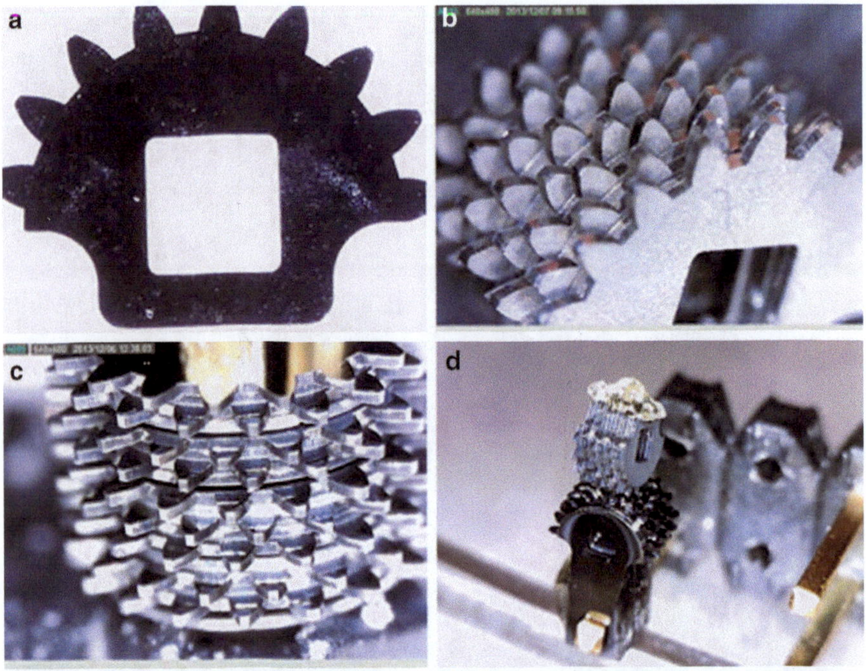

Fig. 7 Micrographs of a millimeter-scale crossed-cylinder wrist mechanism fabricated by stacking carbon nanotube composite sheets. (**a**) A single layer, showing the gear teeth and alignment square for insertion of an electrical pin. (**b**, **c**) An assembled half cylinder. The reflections visible at the top of (**c**) are due to the smoothness of one side of the CNT sheet. (**d**) An assembled wrist mechanism, showing crossed cylinders mounted one on the other

half-cylinder's square) of 3 mm. The mechanism was fabricated from layers of carbon nanotube composite. The material consists of sheets of carbon nanotube forests infiltrated by chemical vapor deposition with carbon. Fabrication details for this process are in Toone et al. (2014). Each sheet was lithographically patterned to produce a half-spur gear with alignment features, as shown in Fig. 7a. Thin spacer layers were also produced to give the right spacing between gear teeth. Because of the resolution used in the layering process (each tooth was produced using one layer), the sloped sides of the rack tooth profile were approximated using a rectangular profile. The layers were approximately 110 μm in thickness. Each half cylinder was then assembled by stacking several sheets in the correct order on an electrical pin. The layers were held together using epoxy spread along the back surface of each half cylinder. Two views of an assembled half cylinder are shown in Fig. 7b, c, and the full wrist is shown in Fig. 7d.

The resolution of the assembly can be improved by making the layers thinner so that the rack tooth profile is sloped rather than approximated as a rectangular profile. To demonstrate this we fabricated layers that were approximately 40 μm thick. This increases the number of layers in each assembly but also improves the

Fig. 8 Photograph of the millimeter-scale crossed-cylinders wrist mechanism prototyped using stacked sheets of carbon nanotube composite material. Each sheet is approximately 40 μm thick

detail and resolution of the part in the direction of stacking. Figure 8 shows one of the half cylinders on an alignment pin prior to the bonding process.

The small-scale mechanism was designed with a pitch diameter of 2.03 mm and 18 teeth on the full circumference. The pressure angle was 25°, and the backlash was 5 %. These small differences from the centimeter-scale prototype are not expected to cause significant performance changes. As with the centimeter-scale prototype, the dedendum of the gear teeth is 1.25 times the addendum.

The wrist is designed to be actuated by thin cables attached to the top half cylinder. Figure 9 shows the arrangement of the actuating cables. One cable is attached along the axis of the top half cylinder to actuate one of the directions of rotation. The other cable is attached perpendicular to the axis of the top half cylinder and is used to actuate the other direction of rotation. The wrist was able to move in both directions of rotation with good range of motion when actuated with the cables.

4 Conclusions

This paper presented a new design for a 2-DOF robotic wrist. The crossed-cylinder wrist mechanism experiences predominately rolling motion, minimizing friction during operation. It is also appropriate for manufacturing at small size scales. We have demonstrated a large-scale model fabricated using FDM 3D printing. The large-scale model validates the smooth motion achievable using the wrist mechanism. It is capable of rotating ±90° in either of two directions (2-DOF). We have further demonstrated fabrication of the wrist mechanism at millimeter scale, with a

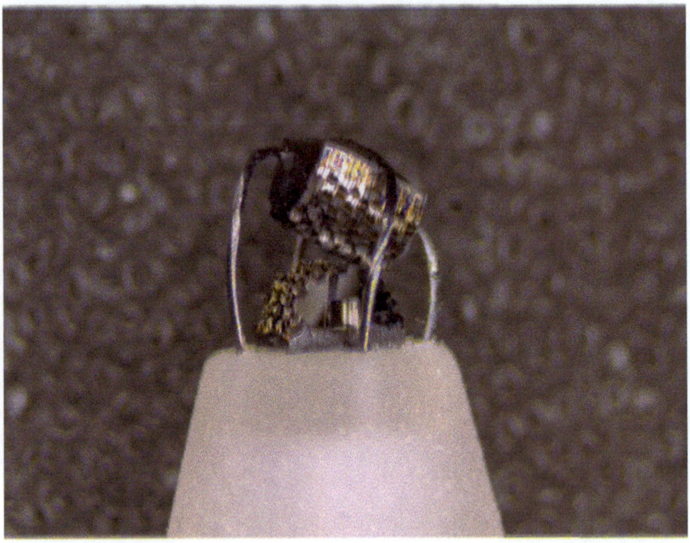

Fig. 9 The actuation cables are arranged orthogonal to each other. Each pair controls one direction of rotation

maximum dimension of 3 mm. The small-scale wrist was fabricated by lithographically patterning sheets of carbon nanotube composite material, and then stacking the sheets to form two geared cylinders. The crossed-cylinder wrist design is expected to be used to improve dexterity and manipulation of small-scale tools, including cutters, graspers, and other robotic tools.

Acknowledgments This research is supported by a grant from Intuitive Surgical, Inc.

References

Halverson PA, Howell LL, Magleby SP (2010) Tension-based multistable compliant rolling-contact elements. Mech Mach Theory 45(2):146–156

Hiller J, Hjelle DA, Lipson H, Nigl F (2011) Bidirectional gear, method, and applications, U.S. Patent no. 21030055838 A1

Howell L, Magleby S, Olsen B (2013) Handbook of compliant mechanisms. Wiley, New York

Jelinek F, Pessers R, Breedveld P (2013) DragonFlex – smart steerable laparoscopic instrument. J Med Devices 7:020911 (2 pages)

Nai TY, Herder JL, Tuijthof GJM (2011) Steerable mechanical joint for high load transmission in minimally invasive instruments. J Med Devices 5:034503 (6 pages)

Shigley JE, Mischke CR (1989) Mechanical engineering design, 5th edn. McGraw-Hill, New York

Toone NC, Fazio WF, Lund JM, Teichert GH, Jensen BD, Burnett SH, Howell LL (2014) Investigation of unique carbon nanotube cell restraint compliant mechanisms. Mech Based Des Struct Mach 42(3):343–354

Zhao B, Nelson CA (2013) Decoupled cable-driven grasper design based on planetary gear theory. J Med Devices 7:020918 (3 pages)
Zoppi M, Sieklicki W, Molfino R (2008) Design of a microrobotic wrist for needle laparoscopic surgery. J Mech Des 130:102306 (8 pages)

Conception of a Mechanical System for Rehabilitation of Hand Function for Use in Medical Training Therapy

M. Feierabend and L. Zentner

Abstract The hands are regarded as the most intensively used parts of the human body. They are used to perform very different activities, and to carry out necessary daily needs. Due to their exposed location and frequent use, hands have a high risk of injury. Given the natural way in which humans use their hands, their major importance only becomes apparent after they are injured. Permanent damage is not limited to physical performance, but can also have psychological effects on the patient. Therefore, the fullest possible rehabilitation of the hand is desirable. Only a very limited number of therapeutically necessary movement tasks can be taken on nowadays using the current technical systems for hand rehabilitation. Motivated by these facts, a mechanotherapy system is conceived, which is fluid-driven and modular. In this paper, the conception of the mechanotherapy system will be limited to one finger. However, the developed solution is transferable to the remaining fingers.

Keywords Compliant mechanism • Precision engineering • Rocker arm • Spring • Spring clip mechanism

1 Introduction

"Man is the wisest of all beings, because he has hands" (Anaxagoras). Early on, through this statement by the Greek philosopher Anaxagoras, passed down by Aristotle in his De partibus animalium, the outstanding and central role of the hands in the human life was clarified.

Precisely because the human hand is such a diverse and highly accurate tool, even small functional deficits can cause a loss of independence and decrease quality of life, in addition to which there can be a negative impact on the social environment. Because of their exposed location and frequent use, hands have a high risk of injury; one third of all injuries and accidents involve the human hands (Schädel-Höpfner and Windolf 2010).

M. Feierabend (✉) • L. Zentner
Department of Mechanical Engineering, Mechanism Technology Group,
Technische Universität Ilmenau, Ilmenau, Germany
e-mail: martin.feierabend@tu-lmenau.de; lena.zentner@tu-ilmenau.de

E.-C. Lovasz et al. (eds.), *Microactuators and Micromechanisms*, Mechanisms and Machine Science 30, DOI 10.1007/978-3-319-15862-4_12

Physicians, occupational and physical therapists have developed a wide range of treatment options that are used for rehabilitation or preservation of hand function after injury or illness. The movement and loading of the injured or diseased muscles, tendons and joints is an important component. This form of therapy is called mechanotherapy. For this, technical devices are increasingly used.

With currently available devices, treatment approaches are implemented, which have mainly been known since the middle of the last century. The technical conditions allow the movement of complete fingers, but the selective moving of individual finger joints is not possible. With the existing devices, both therapists and patients must take into account significant limitations.

In this article, a mechanotherapy system for rehabilitation of hand function is designed for medical use. Here, the basis for a multi-functional system for physiotherapy of the hand will be developed. In addition to the known treatment options, qualitatively new features will be included, making the testing and development of new treatment approaches possible. The goal in using this system is a faster treatment success.

2 Basics

Necessary anatomical terms are clarified in the following chapter. Furthermore possible treatments for diseases of the hand which restrict its mobility are briefly presented. Finally, existing solutions are introduced in the State of the Art and conclusions are drawn regarding the necessary functions of a new mechanotherapy system.

2.1 Anatomical Basics

The skeleton of the hand consists, in addition to smaller sesamoid bones, of 27 bones. As shown in Fig. 1 left, these are divided into various sections. Abutting the arm and the wrist bones (Ochoa et al. 2011; http://www.festo.com/cat/de_de/data/doc_de/PDF/DE/EG_DE.PDF; Kalb 2012) is the metacarpal (Gärtner 2010). The five fingers are connected to the metacarpal. The digits can be divided into four fingers (index to little finger) and the thumb. All fingers consist of a proximal phalanx (Heyer 2010) and distal phalanx. Between these, the phalanx media is located. The bones are connected to joints that are summarized and indicated by the lines in Fig. 1 left. While numbers 1–3 represent the individual components of the wrist, numbers 4–7 show the finger joints, which are also shown in Fig. 1 on the right (No. 1, 2 and 4) (Lippert et al. 2010).

Figure 2 describes the spatial orientation of the hand and illustrates the terms used later for the direction of movement.

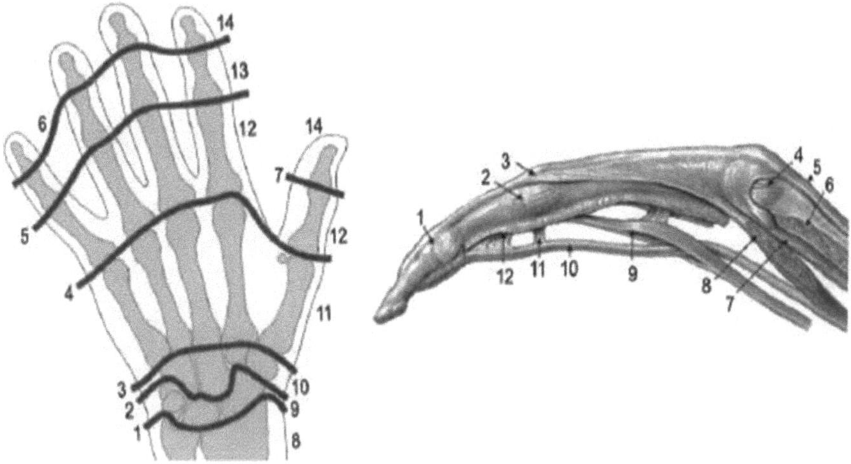

Fig. 1 (*Left*) Structure of the hand, (*right*) joints, muscles and tendons of a finger (Lippert et al. 2010)

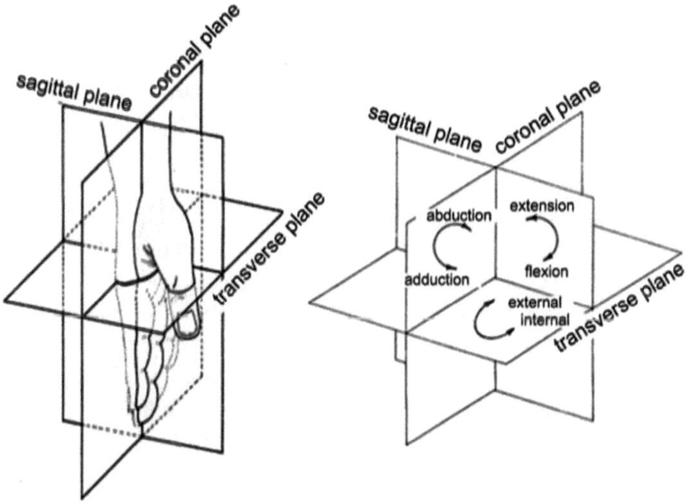

Fig. 2 (*Left*) Principal planes of the hand, (*right*) naming the movements of the hand, depending on the plane and direction of motion (Brand and Hollister 1993)

Each of the four fingers is connected to the metacarpal (cf. Fig. 1 left and right No. 4) by the metacarpophalangeal joint (MCP joint). This joint is able to: spread (abduction), pull (adduction), stretch (extension) and flex (flexion). In contrast, the proximal interphalangeal joint (PIP joint) and distal interphalangeal joint (DIP joint) enable only extension and flexion. The following concepts will be limited to the extension and flexion of one finger. The possible ranges of motion for extension and flexion of a finger are shown in Fig. 3.

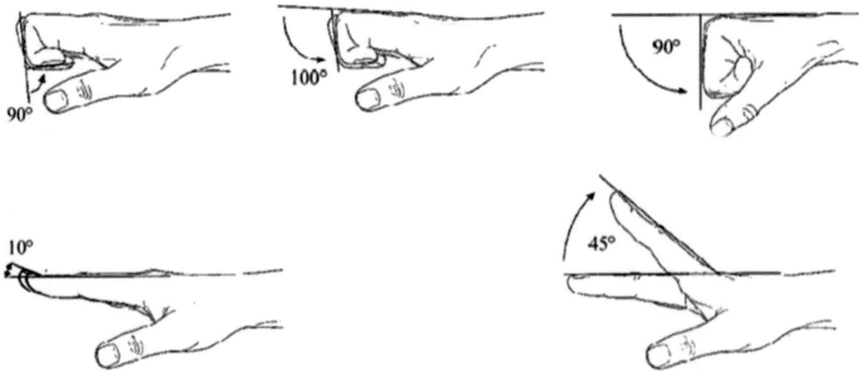

Fig. 3 Mobility of the finger joints; distal interphalangeal joint (*left*), proximal interphalangeal joint (*center*); metacarpophalangeal joint (*right*) (Schünke et al. 2005)

2.2 Treatment Options

For diseases of the hand which restrict its mobility, an extensive rehabilitation program is recommended.

This includes a movement therapy for all impairments, in which the patient exercises either actively (performing with his finger muscles) or passively by an external force (which can mobilize his diseased limbs).

Furthermore, the treatment is mostly supported by splints. Here, static and dynamic ortheses play an important role. Static ortheses restrict the degree of freedom of the joints. Dynamic ortheses inhibit certain movements by means of elastic flexible materials to provide resistance for strength training. Ortheses are also often used in everyday tasks. However, everyday movement tasks can also be specifically imitated and practiced in functional therapy. It is notable that there are customized ortheses for each disease of the hand. They differ mainly according to whether joints are completely fixed, the degree of freedom is restricted or whether a certain movement space is allowed. In order to perform their function, ortheses are usually customized to fit the hand of the patient exactly.

If a passive motion therapy is recommended for the patient, it generally requires a second person, if not a therapeutic specialist who practices with the patient. This imposes an even greater dependence on the already limited autonomy of the patients, but also represents a social problem. The growing burden of health insurance is just one example.

2.3 State of the Art

The existing solutions are analyzed, prior to the conception of a new mechanotherapy device.

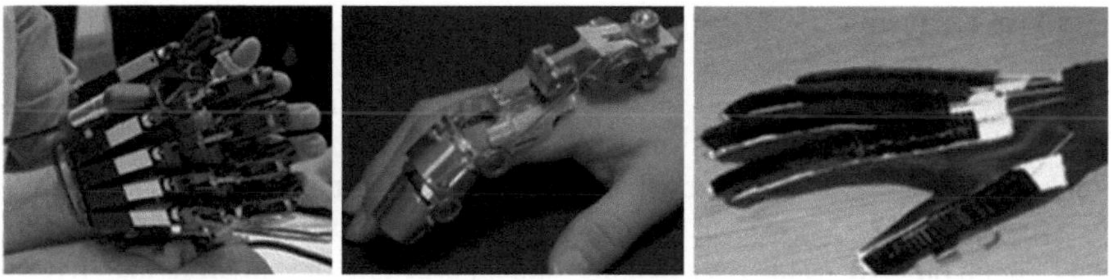

Fig. 4 Mechanotherapy systems (*left*) Hand of Hope from Rehab Robotics (Ho et al. 2011), (*center*) prototype of the Handexos for the index finger (Chiri et al. 2011) (*right*) J-Glove with Bowden cables on the back of the hand (Ochoa et al. 2011)

2.3.1 Hand of Hope

The Hand of Hope is distributed by Rehab Robotics and was developed at the Technical University of Hong Kong. The device (Fig. 4 left) consists of five individual modules, which are each connected by a linkage to a linear electric actuator with an integrated control unit on the back of the hand. Via curve gears, the forces of the actuators are converted into a moment. One module can mobilize the metacarpophalangeal and the proximal interphalangeal joint within a maximum angle of 65° (for stretching and bending). Independent movement of the two joints is impossible (Ho et al. 2011).

2.3.2 Handexos

In 2009, a research team led by Azzurra Chiri, presented a mechanotherapy system called Handexos. It is an exoskeleton made of aluminum in a shell construction which is worn on the hand (Fig. 4 center). This has five modules. Each module is driven by a DC motor. Bowden cables that connect the motor with a small deflection wheel located at the joints set the torque of the motor into a stretching motion of the hand. Return springs also connected with a Bowden cable move the Handexos back to the starting position. However, no single joints can be controlled (Chiri et al. 2011).

2.3.3 J-Glove

In 2011, Jose M. Ochoa presented his research results for a therapeutic glove called the J-Glove. A Bowden cable is located on the top of each finger. These five cables are united on the forearm to a main pull (Fig. 4 right). A DC motor, which is connected to the main pull, moves the Bowden cables to the front or to the rear, thus forcing all of the fingers to a flexion or extension (Ochoa et al. 2011).

2.3.4 Conclusion

A variety of technical supports for the rehabilitation and treatment of hand diseases exists. However, they are not multifunctional or portable and do not allow the individual to receive separate treatment of individual joints of the fingers.

No mechanotherapy system that can be modularly adapted to the size of a patient and the therapeutic needs (targeted movement, blocking or inhibiting the affected joints) yet exists, but such a system could positively influence the rehabilitation process. Electively, an actuator, a locking element or an inhibitor should be integrated for each joint of a finger. An actuator allows (due to its force generation) the joint to perform a passive movement. An inhibitor represents a surmountable resistance for independent motion. A locking element fixes a particular joint at a certain angle.

3 Conception

In the following chapter the development of the mechanotherapy system is described. This is inspired by the constructive development process, and includes the specification of the development goal, the principle of synthesis for actuator, actuator position, and the mechanism. Finally, the complete system is introduced briefly.

3.1 Clarification of the Task

The goal of the development is the passive movement of the finger joints in their full physiological motion space. Furthermore, active movement should be possible. For this feature, actuators are utilized, allowing a passive movement of the finger and offering a defined resistance against finger movement during an active movement. In addition, the use of passive elements is conceivable. They consist either of a rigid body to lock the joint or a compliance that allows the movements against a resistance respectively to apply a uniform force.

For the drive, pneumatic actuators are preferable as the damping properties of the air mean that the drive is shock-free and jolt-free. Power transmission by wire ropes or Bowden cables is excluded from the beginning, as these require large structures and are a potential hazard to patients.

The force transmission takes place only in the elements adjacent to the joint (bones) and the surrounding tissue. The neighboring joints are not included, in order to move all joints independently. Selective control of each joint should be possible. It must also be ensured that no incorrect loading of joints occurs through the system.

In addition, a lightweight build and a space-saving design with a modular construction should be achieved.

In the design process, the main focus is on the actuator concepts, the positioning of the actuator and the mechanism for the transformation of the actuator force into a moment which moves the phalanx of one finger.

3.2 *Principle Synthesis*

3.2.1 Actuator Concepts

A brief overview of possible pneumatic actuator concepts is given below. For actuators, commercially available pneumatic drives or new self-developed assemblies are suitable. Figure 5 shows an overview of possible actuators. The solutions can be divided into three main groups: cylinder, pillows and bellows.

Cylinder

Even the smallest cylinders are relatively large. The shortest pneumatic cylinder identified has an extended length of 34 mm (http://www.festo.com/cat/de_de/data/doc_de/PDF/DE/EG_DE.PDF), which is difficult to attach to the phalanx of a finger. Furthermore the generated force as well as the stroke is insufficient. But self-designed cylinders with integrated compliants (by use of elastomers) are promising in terms of stroke and force.

Bellows

Another way to drive the pneumatic system is with bellows made of elastomers. The size of the actuator can be aligned to the required dimensions. An example of a customized structure is a bellows with only one fold. This can be designed to be much more compact than types with several folds. Rotationally symmetric bellows unfold when their inner pressure is increased. The result is a purely translational motion. If an asymmetrical bellows is designed, it moves along a curved path. Such asymmetric structures have a certain potential for rotary pneumatic actuators.

Cushion

Rubber or rubber-like soft materials are characterized, among other things, by their hyperelasticity. Therefore, an actuator which is similar to a balloon is feasible. Increasing the inner pressure of the balloon causes it to expand nondirectionally.

basic shape	movement type	effective direction	constructive details	sketch pressure p_1	sketch pressure $p_2 > p_1$
cylinder	translational	single or double-acting with compliant			
cylinder	translational	single or double-acting			
bellow	translatorisch	single-acting	one fold		
bellow	translatorisch	single-acting	several folds		
bellow	rotatorisch	single-acting	asymmetrical		
cushion	nondirectional		no special surface structure		

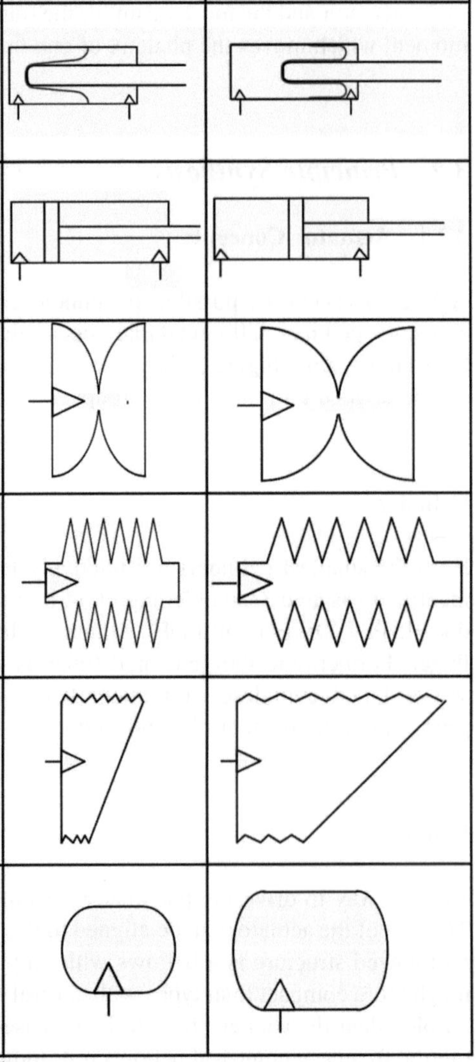

legend: air supply and exhaust

Fig. 5 Table presenting various pneumatic actuator concepts

3.2.2 Actuator Position

The actuator can be deployed either directly, in the area surrounding the joint or adjacent to the finger, for example, on the metacarpus, the forearm or a body distal object such as a table. An actuator directly on top of the finger can extend and contract along three coordinates: up and down (dorsopalmar), along the finger (longitudinal), or laterally (transverse).

3.2.3 Concepts for the Mechanism

In order to transmit the force or torque of an actuator to a finger joint, a mechanism is necessary. In addition, the actuator also has the function of converting the stroke of a translational drive into a rotation.

The force transmission is to be achieved directly through the system attached gear and joint elements. When selecting a mechanism, a pure torque should act on the finger joint about the rotation axis of the finger. Furthermore, no longitudinal and vertical forces should appear. If this nevertheless occurs, the finger joint could be seriously injured.

To minimize disturbing forces on the finger joint, the joint of the mechanotherapy system has to be in line with the finger joint. To ensure this, the joints of the mechanotherapy system have to be sidewise to the finger joint. Furthermore, it should be taken into account that the MCP joints are not accessible due to their linkage to the metacarpus. A cam mechanism avoids this problem by guiding the movement along a circular path, where the virtual axis coincides with the rotation axis of the finger joint.

Figure 6 provides an overview of mechanisms that could transfer the actuator force to the joints. A force is applied by the schematically shown actuator, which results in an angular change.

The scotch-yoke mechanism is a classic four-bar mechanism. Within a cam mechanism, the real rotation axis is replaced by a circular arc guide. Through a gearwheel, the rack transmits either the force of the actuator into a torque or the stroke of the actuator into a rotation. Due to the anatomical conditions, the gearwheel can only be attached sidewise to the finger. The "scissors" is the only model that requires an actuator, with a circular path operating force.

3.2.4 Final Version for a Mechanotherapy System

After several principle solutions for the actuator, the actuator position and the mechanism were outlined, and these were combined with the technical principles and evaluated (compare (Kalb 2012)). In the final version of the mechanotherapy system, the PIP and DIP joints are mobilized through a cam mechanism with aligned axis positions (compare Fig. 7). The mechanism was designed for a large continuous motion range, and an efficient power transmission.

Since the MCP joints are not freely accessible because of their link to the metacarpal, no rotation axis of a mechanism can align with the axis of the finger joint. Therefore, a cam mechanism without Alignment is used (compare Fig. 8).

Axis position	Gear designation	Sketch
Aligned	Scotch-yoke mechanism	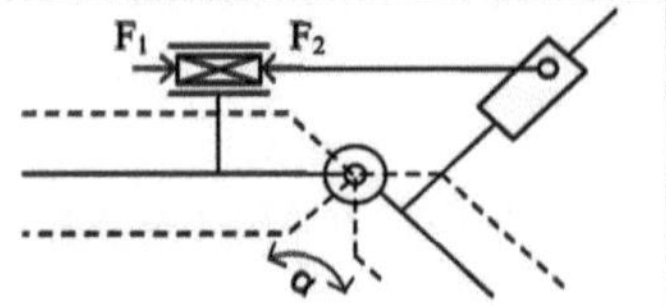
	Rack mechanism	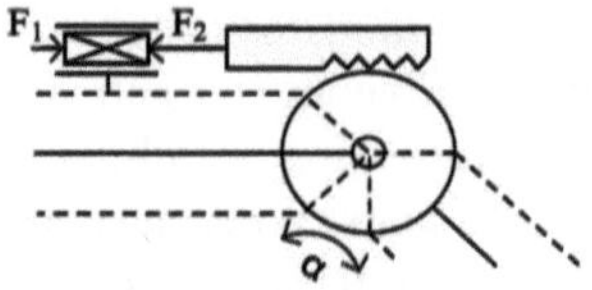
	"Scissors"	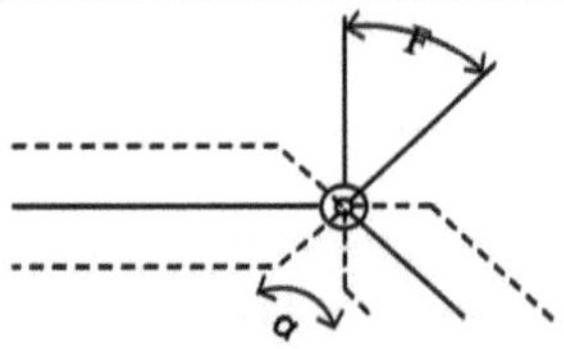
Virtual aligned	Cam mechanism	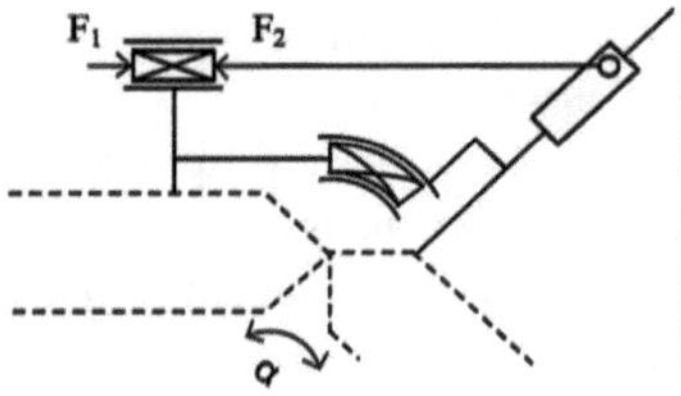

Fig. 6 Table presenting different mechanisms for the transformation of the actuator force into a torque

Three actuators are used for each finger. These are positioned directly on top of the fingers (longitudinally) to generate the forces necessary for the movement. Through a right designed cylinder both high forces and a large stroke can be achieved. For this reason, self-designed cylinders with integrated compliants are used as actuator. The joints can be moved by approximately 65°. The range of motion can be traversed continuously. A selected movement of individual joints is possible. The indicated actuators can be replaced by passive elements, which will inhibit or prevent active movement of the fingers.

The rigid frame structure simplifies the mounting of actuators and sensors. It is also possible to integrate a size adjustment for different hand sizes. By padding the

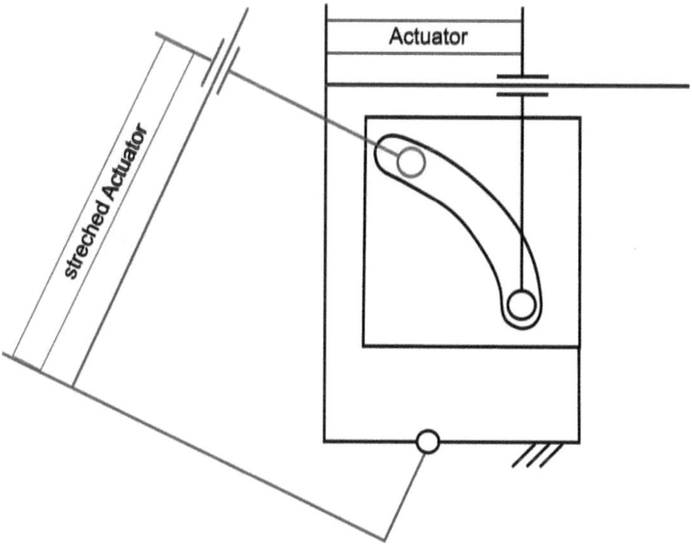

Fig. 7 Principle picture of Mechanism for PIP and DIP joints

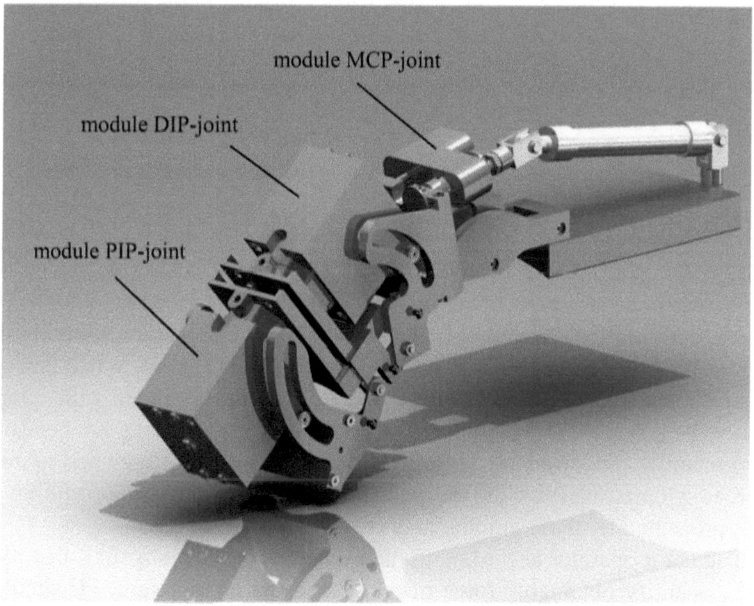

Fig. 8 CAD model of the existing prototype

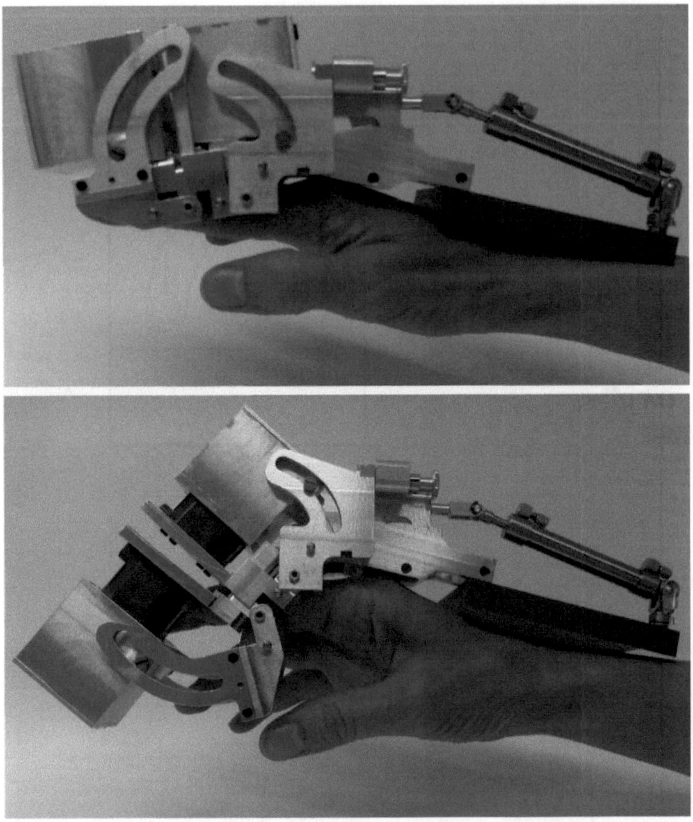

Fig. 9 Movement of prototype from a stretched into a strongly flexed position

inside of the frame structure, an individual adaptation to various hand forms (finger diameter, deformation) is possible. A prototype was manufactured (see Fig. 9).

Figure 10 shows actuator assembly for the PIP and DIP joint. When flexion and extension are to be performed, either a double acting cylinder with integrated compliant or a single acting cylinder with compliant and a spring are needed. The driving members are arranged to ensure the largest possible stroke and to transmit the forces symmetrically on the mechanism. Furthermore, twisting or tilting of the slide is prevented by three guiding areas.

The actuator operates according to the principles of unrolling and stretching. In order to quantify obtainable force in relation to the air pressure, FE simulations were performed. The result of these studies is an actuator with a height of 17 mm, a diameter of 10.9 mm at the thickest spot and a wall thickness of 1 mm (compare Fig. 11).

The simulation was 2D axisymmetric with hyperelasticity. As material the silicone M4644 was used with a Shore hardness of 40. The necessary material data for the simulation model was fitted on the basis of uniaxial tensile data.

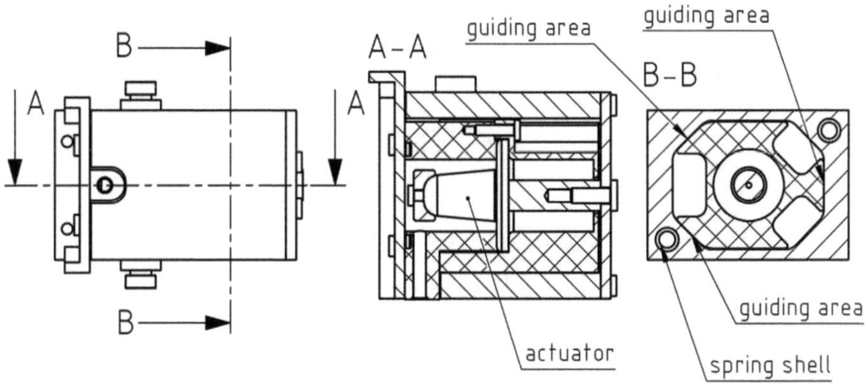

Fig. 10 (*Left*) actuator assembly PIP and DIP joint (*center*) longitudinal section (*right*) cross-section

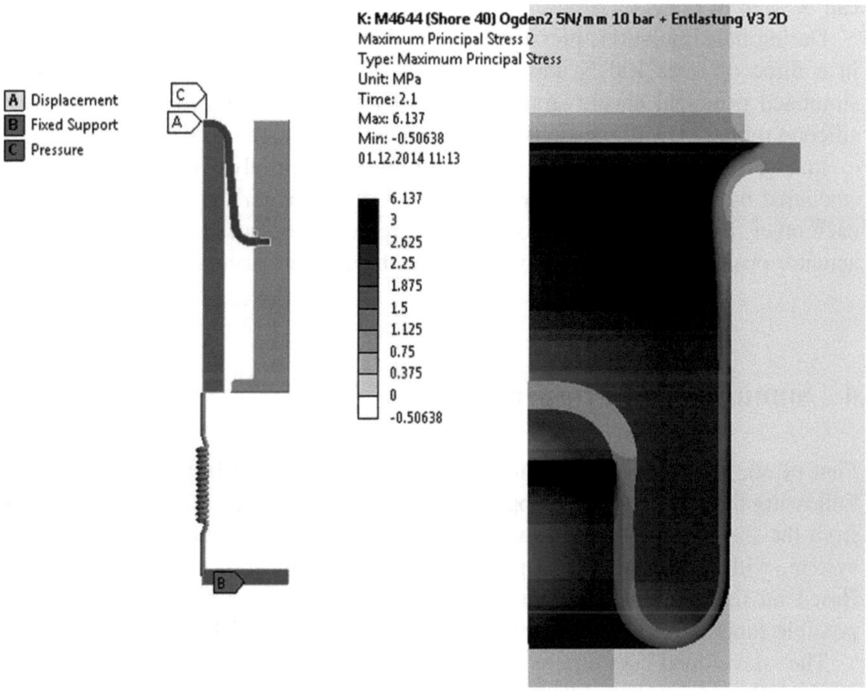

Fig. 11 (*Left*) Model of actuators with boundary conditions and loads; the counterforce of the diseased joint was modeled by a spring, the pressure was applied on the upper curved surface (*right*) simulation of the actuator using Ansys Workbench maximum principal elastic strain by an applied pressure of 1 MPa

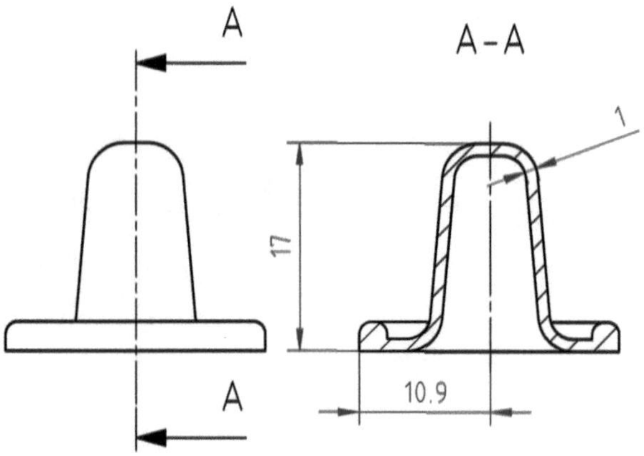

Fig. 12 Dimension of actuator designed by FEM

Meshing was performed with Plane183 elements (8 nodes). The simulation setup can be seen in Fig. 11.

During the simulations pressure up to 1 MPa could be achieved, which resulted in a force of over 100 N and a stroke of 30 mm. Furthermore the maximum simulated principal elastic strain was smaller than the tear strength of the used silicone rubber. The dimensions of the used actuator can be seen in Fig. 12.

In Fig. 13, some simulation results are presented briefly. The greater the load (modeled by springs with different stiffness), the stronger the curves differ from each other. By measuring the volume and pressure, it is possible to determine the actuator position and actuator force, both with sufficient accuracy.

4 Summary and Prospects

First of all, the basis of the anatomy and physiology of the hand was introduced. Following this, the technical supports used for therapy were considered. Resulting from the shortcomings of known therapeutic devices, the requirements for a new system, with particular emphasis on orthopedic technical aspects, were situated. Based on these preliminary considerations, the tasks were clarified and several possible function principles for the new system were established.

The developed system is characterized, in comparison to the known mechanotherapy systems, through a variety of special functions. It is ideal for passive movement of the hand as well as for dynamic ortheses during treatment, with individually adjustable, mechanical properties. The system is modular, so simultaneous or separate treatment of the finger joints can be performed. The possibility of a subsequent integration of sensors opens the way for an objective therapy evaluation (measurement of the enlargement of the movable range and the actual load during the treatment). Due to the above-mentioned characteristics of the system, an improvement of the treatment effectiveness is expected.

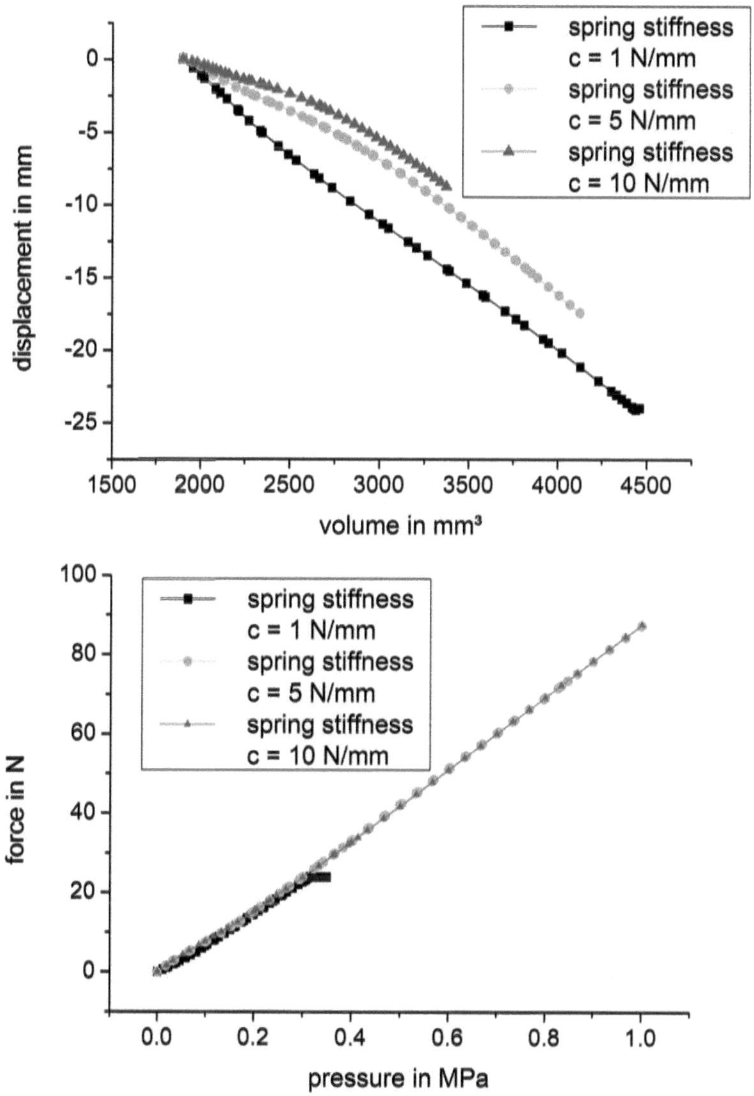

Fig. 13 Results of the FE-simulation at different counterforces (*above*) volume over pressure (*below*) pressure over force

References

Anaxagoras Überliefert durch Aristoteles, De partibus animalium, IV, 10;687 a 7
Brand PW, Hollister A (1993) Clinical mechanics of the hand. C.V. Mosby, St Louis
Chiri A, Vitiello N, Giovacchini F et al (2011) Mechatronic design and characterization of the index finger module of a hand exoskeleton for post-stroke rehabilitation. IEEE/ASME Trans Mech 17:884–894

Gärtner R (2010) Entwicklung eines Mechanotherapiesystems für die menschliche Hand. Diploma dissertation, Technical University, Ilmenau

Heyer T (2010) Konzeption eines pneumatisch aktuierten Handschuhs für die Mechanotherapie der menschlichen Hand. Diploma dissertation, Technical University, Ilmenau

Ho NSK, Tong KY et al (2011) An EMG-driven exoskeleton hand robotic training device on chronic stroke subjects: task training system for stroke rehabilitation. In: Rehabilitation robotics 2011, IEEE International Conference on Rehabilitation Robotics (ICORR), IEEE, Zurich, Switzerland

http://www.festo.com/cat/de_de/data/doc_de/PDF/DE/EG_DE.PDF. 12th July 2012

Kalb J (2012) Entwicklung eines pneumatisch aktuierten Systems für Mechanotherapie der Hand. Bachelor dissertation, Technical University, Ilmenau

Lippert H, Herbold D, Lippert-Burmester W (2010) Anatomie. Text und Atlas; deutsche und lateinische Bezeichnungen. Elsevier, Urban & Fischer, München

Ochoa JM, Listenberger M et al (2011) Use of an electromyographically driven hand orthosis for training after stroke. In: Rehabilitation robotics 2011, IEEE International Conference on Rehabilitation Robotics (ICORR), IEEE, Zurich, Switzerland

Schädel-Höpfner M, Windolf J (2010) Handchirurgie in der Unfallchirurgie. Unfallchirurg 113:10

Schünke M, Schulte E, Schumacher U, Voll M, Wesker K (2005) Prometheus LernAtlas der Anatomie. Georg Thieme, Stuttgart

Modeling, Parametric Synthesis and Characterization of Bellow-Type Pneumatic Micro-actuator

J. Prateek and G. Prasanna

Abstract Pneumatic micro-actuators present unique advantages such as high force to volume ratio, tunable to get desired force vs displacement characteristics, free from high voltages, to list a few. Among several types of pneumatic actuators bellow type actuators presented recently are more conducive for MEMS fabrication. This paper presents extensive parametric analysis of these actuators to finally generate design and synthesis curves in terms of non-dimensional parameters. Analysis is carried out with finite element method (FEM); based on results and physical insights, non-dimensional parameter for synthesis are identified. Results are presented as design curves in non-dimensional parameters in such a way that based on desired force displacement curve users can converge on parameters without a need for further FEM for their specific case. Results show interesting possibilities of large variety of desired "force-displacement curves" spanning feasible range of parameters. One specific case is considered for fabrication of this bellow type pneumatic actuator and its characterization validates the proposed design analysis.

Keywords Pneumatic Micro-Actuator • ANSYS

1 Introduction

The discipline of MEMS has grown quite vast and multiplying very fast. The advancement in fabrication of Micro-Nano Electronics, semiconductor devices and Micro-fabrication techniques, has laid very strong foundation for MEMS to evolve. The idea for developing pressure based micro-actuator arises from the application of Micro-handling devices like micro-grippers, micro-surgical devices, catheter, forceps (De Greef et al. 2009) etc. Pneumatic mechanical micro-actuators when compared with other actuators like piezoelectric, electrostatic etc. are far superior in terms of force per unit volume, high blocking force, and displacements simultaneously (De Volder and Reynaerts 2010). The motivation was to develop a

J. Prateek (✉) • G. Prasanna
Department of Mechanical Engineering, Indian Institute of Technology Bombay, Mumbai, India
e-mail: prateek.urs@gmail.com; gandhi@iitb.ac.in

E.-C. Lovasz et al. (eds.), *Microactuators and Micromechanisms*, Mechanisms and Machine Science 30, DOI 10.1007/978-3-319-15862-4_13

novel pressure based micro-actuator for wide range of blocking forces and displacements. The main advantage of bellow-type actuator over membrane type is longer linear travel for the similar dimensions and high forces (Yang et al. 1997).

The concept of bellow type pneumatic actuator was presented first in (Butefisch et al. 2002) where flexure mechanism for micro-gripper was actuated. The modified design was presented in (Gandhi et al. 2011a) where piston was precisely guided using parallelogram flexure mechanism to avoid jamming. This paper provides design curves and procedure for designing such actuators given the actuator specifications in terms of blocking force and free displacement. Extensive FEM analysis is carried out in ANSYS and various critical parameters identified, pertaining to the analysis, are plotted. Study and analysis shows that, wide range of desired "Blocking force versus Displacement" characteristic curve can be obtained by appropriate choice of design parameters. The key feature of the proposed parametric synthesis is, "a unique parameter R to design a desired pneumatic micro-actuator, with required blocking forces and displacements". Using the procedure proposed in this paper, one actuator is fabricated using SOI MUMPS process. The device is packaged and experimental verification of characteristic curves is plotted for this specific case.

This paper is organized as follows: Section II presents actuator modeling and parametric analysis of the system based on extensive FE simulations. Section III highlights the actuator fabrication using SOI-MUMPS. Section IV highlights the packaging and experiment for a special case. Finally Section V concludes the research findings followed by acknowledgements.

2 Design, Modeling and Parametric Synthesis

This section first presents the proposed design of the bellow actuators and highlights the enhancements incorporated in the design as compared to the previous. Next, the results of extensive FE analysis carried out to simulate the displacement and blocking force produced by applying pressure to the actuator are presented. Simulations are carried out for parametric analysis and results are presented in a non-dimensional form that can be easily used for synthesis of other such micro-actuator with the desired design specification.

2.1 Design

The basic structure of the micro-pneumatic actuator, as shown in Fig. 1 consists of a piston connected to housing by two spring elements. This design is similar to that of earlier designs, presented in literature except for the flexure guides. Symmetric bending of the lobes under this uniform load allows the piston to move in vertical y direction. The motion of the output piston normally is along one y direction in the

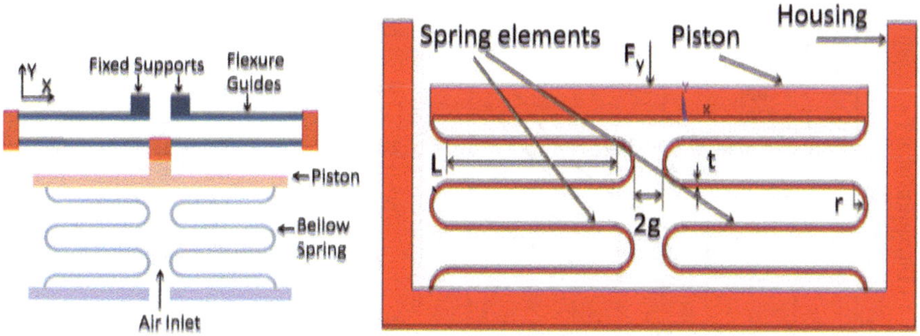

Fig. 1 Proposed design of bellow actuator with guide flexure and actuator parameters

absence of external force in x direction or moment around z axis. However, if an external device to which the actuator is connected exerts force in x direction (or moment about z), a lower stiffness in that direction would disturb the desired function of the actuator. To overcome this problem a standard parallelogram flexure guide is added to the actuator as shown in Fig. 1. The flexure guide makes sure that the stiffness of actuator is much higher in all directions other than the desired motion. Thus parasitic motions in other directions would be prevented and the precision & repeatability of actuator motion would be enhanced.

2.2 *Analysis*

Various parameters affecting performance of actuator include length (l), radius (r), thickness (t) of spring element, gap (g) between the two spring elements and number of lobes (n). Figure 1 displays these parameters for actuator with one lobe. To develop compact actuator (overall length not more than 1,200 μm), and keeping in mind constraints of the SOI MUMPS process (fabrication process), nominal values of the parameters chosen for design analysis are as follows: l = 300 μm, t = 5 μm, r = 30 μm, g = 25 μm and n = 1. The nominal design with parameters mentioned above is modeled and analyzed by Finite Element Method using ANSYS. The structure is modeled using PLANE 82 element with "Plane stress with thickness". The model is meshed using smart size, which ensures finer meshing at places with smaller dimensions and different element size with change in dimensions for whole device. The applied boundary conditions are, uniform internal pressure along the springs, blocking force at the central node of piston, and zero displacement at the spring base.

Simulation in ANSYS was performed with change in the number of elements ensuring the required convergence of analysis. The results of simulations showed approximately 27 μm of displacement of piston in y-direction and maximum Von-Misses stress is 0.21 GPa. This displacement is much higher than that achievable by other actuators such as comb-drives of the comparable dimensions and the stresses

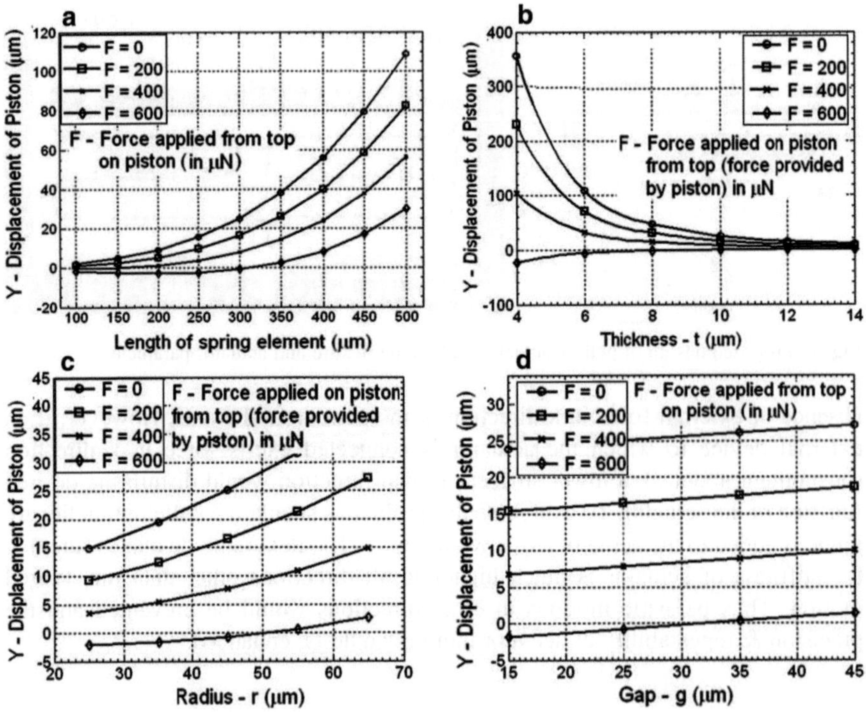

Fig. 2 Variation of displacement of piston in Y – direction with change in length (**a**), thickness (**b**), radius (**c**) and gap of spring element (**d**), and force applied from top at constant input pressure of 60 kPa

are much lower than stress limit of silicon. Thus for such actuators considering current fabrication limitations, the strength as limiting factor will seldom be an issue. Along with displacement, output force provided by the piston (blocking force) is another quantity of interest. Blocking force would be governed in general by the application needs. For example, the actuator placed at the input of micro-flexural logic gates in (Modi et al. 2007), should give enough force to displace the gate and produce meaningful displacement corresponding to the desired logic. Hence, to study effect of this force on the actuator performance in simulation, we consider a force F_y in the opposite direction to the actuator displacement as shown in Fig. 1. Simulations are carried out for constant input pressure and different values of force F_y and the results are presented in Fig. 2.

The results (with $n = 1$) show that displacement increases in quadratic and cubic manner when length and thickness of link is varied (Fig. 2a, c). The effect can be justified from the basic cantilever deflection using Euler beam bending theory. Similarly deflection increases linearly with increase in radius (Fig. 2b), but unaltered when gap, g between the lobes is altered (Fig. 2d).

The simulations are carried out for a range of parameters and finally, it is concluded that

1. The proposed actuator shows promising results to produce the desired high blocking force and high displacement
2. The most effective parameters are L, length of lobe, t, thickness of the lobe and R, radius of lobe.

The next subsection generalizes these results using these parameters primarily L and t. And plot the curves for different R.

2.3 Parametric Synthesis

This subsection extends the analysis further to enable ease in selection of design parameters of actuator based on user defined performance. Actuator design starts with first arriving at the characteristic curve of the actuator, based on application requirement and find the slope of the curve (points in between typically follow a straight line with slope corresponding to "actuator stiffness"). For example, Fig. 3a shows blocked force vs. displacement variation to be a straight line; the slope of this line is the actuator stiffness. We analyze and present effect of design parameters L; r; and t (assuming $n = 1$) on actuator stiffness and "maximum blocking force" as outputs. Since there are several design parameters, it would be useful to identify combination of them that can simplify the design process of actuator and get better insight into the design space. After several iterations and deliberation, it is observed that when these outputs are plotted against a ratio, defined as:

$$R = \frac{t^3}{L^{2.1}} \tag{1}$$

yield a straight line variation. Hence, this ratio is chosen for plotting the output characteristics. Choice of the combination is motivated from Euler-beam bending theory applied to cantilever with uniform loading. A set of thickness, viz. [06 08 10 12 14] μm, set of length viz. [200 250 300 350 400 450 500 550 600] μm and set of radius, viz. [25 35 45] m, are defined to carry out the simulations to obtain output variations against the ratio defined above. The above sets are chosen such that the overall actuator dimensions span orders of microns to a few millimeters. Approximately 400 simulations are carried out to consider all combinations of parameters possible from the above mentioned sets. From these simulations outputs, namely actuator stiffness and maximum blocking force, are captured and plotted against the proposed ratio R for various values of the parameter radius (note that several combinations of t, L and r may yield the same R).

Figure 3b, c shows the variation of maximum blocking force and actuator stiffness with R at pressure of 60 kPa for different radius. It can be observed from Fig. 3c that the maximum blocking force is independent of radius of the lobes. Thus with the proposed ratio, design of actuator simplifies, as follows:

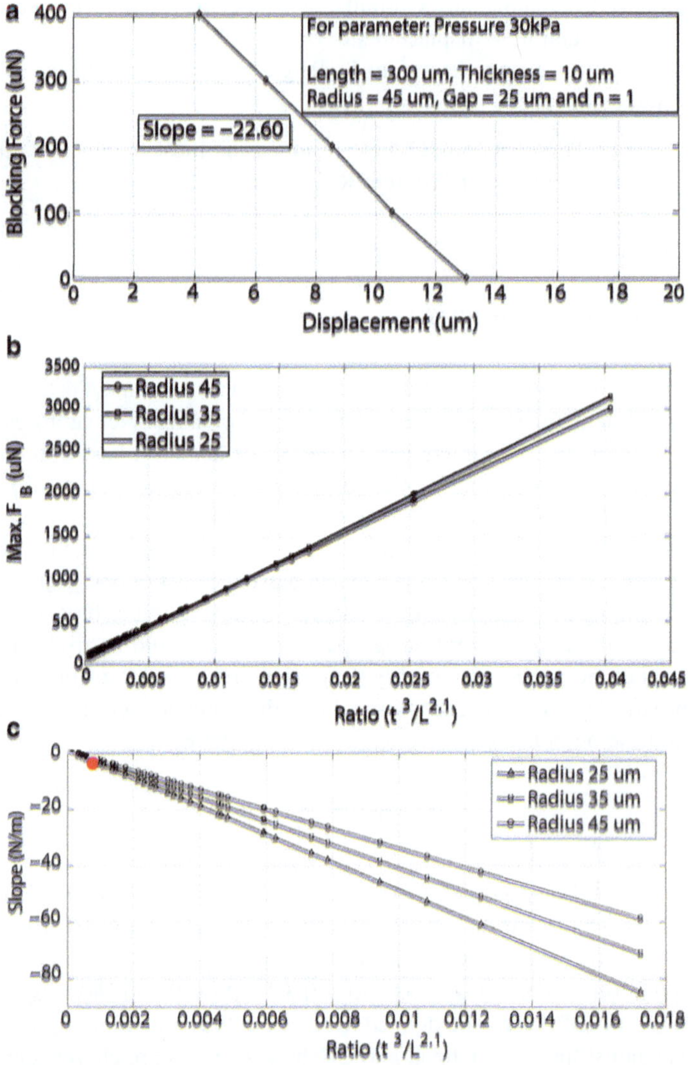

Fig. 3 (**a**) Characteristic curve for pneumatic micro-actuator (for example), (**b**) Various Curves relating change in maximum blocking force with ratio R for different radius, (**c**) Various curves relating change in slope of with ratio R for different radius

1. Given the maximum blocking force to be produced by actuator, choose R from Fig. 3b.
2. For the chosen R, depending on either stiffness or maximum displacement (as shown in Fig. 3a) required, choose radius using Fig. 3c.
3. Once R is known, combinations of lengths and thicknesses can be explored for design depending on space constraints and fabrication possibilities.

It is observed that, these graphs scale proportional to the Young's modulus of material. So, for new material (E1 Young's modulus), the actuator stiffness should be multiplied by E1/190 and new curves corresponding to Fig. 3b, c, can be obtained. (190 GPa is Young's Modulus of Silicon (Peterson 1982), used in our analysis).

3 Actuator Fabrication

Considering strength and fabrication capabilities, silicon is selected as structural material and SOI wafer for development. The fabrication process SOI MUMPS (Miller et al. 2004) is chosen for developing initial prototypes because it offers several advantages:

1. No sealing from bottom is required as used in (Butefisch et al. 2002) for developing these actuators. The bottom Si layer would act as sealing from below.
2. This process can be tuned to release structures without bulk etching.
3. It offers ease and economy of fabrication.
4. Further scaling down of actuators can be possible with SOI processes.

Figure 4a shows the schematic design of actuator developed using LEDIT (Miller et al. 2004). A flexure guide mechanism is incorporated at the piston end to ensure that the lobes are properly guided and they do not touch the sidewalls. Figure 4b shows, SEM of actual fabricated structures using SOI MUMPs process. The structures clearly show piston, lobes and flexure mechanism. These structures were further processed for packaging and characterization of actuators is carried out.

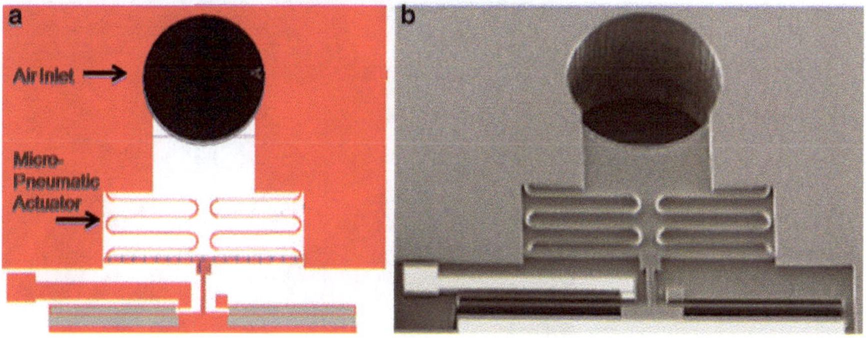

Fig. 4 Pneumatic micro-actuator (**a**) Designed in LEDIT and (**b**) SEM image

4 Packaging and Experiments

Packaging first started with developing tubing connections to the actuator. The tubing connections are permanently made to an acrylic plate to match with SOI chip holes for actuators. Further a frame is developed around these holes to act as locating surfaces for SOI chip (see Fig. 5). Once SOI chip is located using frames, the holes in acrylic plate match with those in the chip. A specially cut glass piece (held in place by frame and top metallic cover) covers the actuator portion only and part of piston is accessible from outside. In this way multiple chips can be characterized by easy assembly and disassembly.

For characterizing the blocking force, a probe is used for applying known blocking force in the opposite direction (at piston) using custom-built mechanical probe station (Gandhi et al. 2011b) (see Fig. 5). An optical microscope is used to visualize and measure displacement of the actuators. Figure 6 shows schematic of pneumatic circuit used. Festo flow control valve is used to control the pneumatic actuator and pressure applied to the piston is sensed using a STM pressure sensor.

The force is applied so as to nullify any displacement of the piston under the actuation pressure controlled by pressure regulator. First, the displacement of actuator is determined by varying applied pressure. Figure 7a shows the variation of blocking force with Displacement, against the applied pressure. The experimental and simulation comparison of pressure vs. displacement is presented at 60 kPa. Further experiments can be done to validate the parametric synthesis.

Fig. 5 Packaging of SOI wafer with pneumatic connections for experimental characterization

Fig. 6 Schematic of pneumatic circuit used for actuation

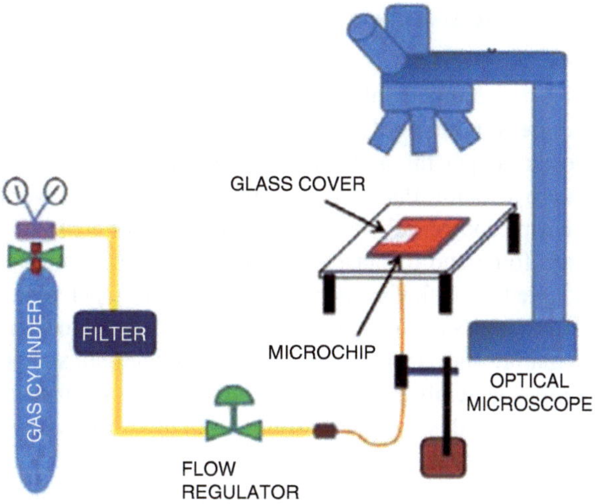

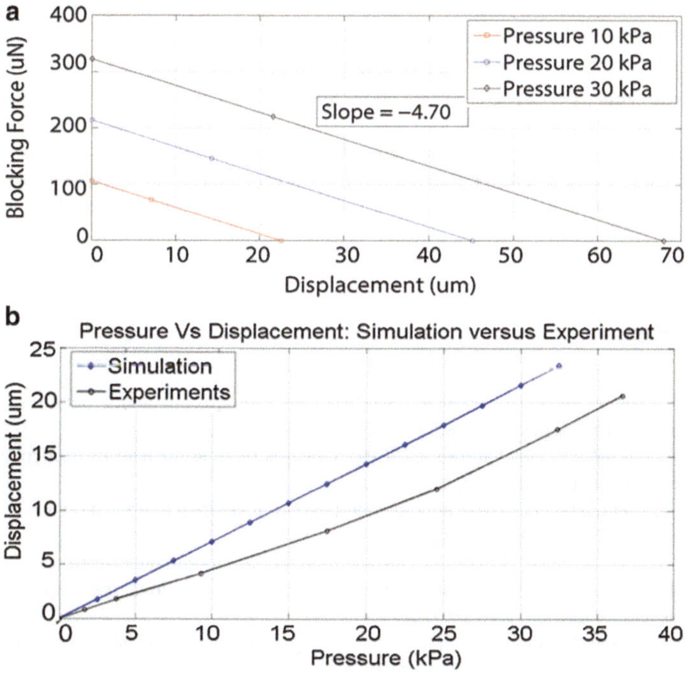

Fig. 7 (**a**) Characteristic curve for the actuator for different pressures, (**b**) Comparison of experiment with simulations at 60 kPa

5 Conclusions

The development of micro-actuators is rapidly evolving and continuously new progress in terms of efficiency, power, and force output is reported. Flexibility in design as per the application requirements is essential characteristics of an ideal micro actuator. Pneumatic micro-actuators have several advantages as compared to their piezoelectric, electrostatic, thermal counterparts: high force per unit volume, high force and displacements possible at the same time, and high energy density coupled with flexibility is tuning design as per requirements. This work demonstrates the said advantages for the enhanced design of bellow type pneumatic actuator. Based on the proposed characteristics ratio R defined in this work, a simple generalized design procedure is evolved using results of extensive finite element analysis (FEA) carried out in this work. Actuator is fabricated with silicon as the material using SOIMUMPs process technology. Further packaging and characterization of the actuator in terms of free displacement and blocking force is carried out using a novel custom-built 'mechanical probe station'. Overall the paper evolves a simple design procedure, demonstrates flexibility and effectiveness of pneumatic actuator for range of actuator needs for MEMS.

Acknowledgments This work is carried out at Suman Mashruwala Advance Micro-engineering Laboratory (http://www.me.iitb.ac.in/~mems) setup with generous donations by Mr. Raj Mashruwala. Authors also acknowledge contribution to this work by Mrs. Himani Shah Modi and Mr Anand Savalia in getting SOI-MUMPS fabrication and initial preparation for characterization. The CNC drilling was carried out at Machine tools laboratory with the help of Mr Pratik Mahambre.

References

Butefisch S, Seidemann V, Büttgenbach S (2002) Novel micro-pneumatic actuator for MEMS. Sensors Actuators A Phys 97–98:638–645

De Greef A, Lambert P, Delchambre A (2009) Towards flexible medical instruments: review of flexible fluidic actuators. Precis Eng 33:311–321

De Volder M, Reynaerts D (2010) Pneumatic and hydraulic microactuators: a review. J Micomech Microeng 20:18

Gandhi P, Savalia A, Shah H (2011a) Design, fabrication, and characterization of a pneumatic micro-actuator. In: Proceedings of IMECE 2011, ASME international mechanical engineering congress and exposition, Denver, CO

Gandhi P, Prajapati V, Shah H, Pai R (2011b) Development of mechanical probe station for characterization of micro-flexure mechanisms. In: Proceedings of IMECE 2011 ASME international mechanical engineering congress and exposition, Denver, CO

Miller K, Cowen A, Hames G, Hardy B (2004) SOIMUMPs design handbook. MEMScAP. http://www.memscap.com/__data/assets/pdf_file/0019/1774/SOIMUMPs.dr.v8.0.pdf

Modi A, Gandhi P, Shah H, Singh S (2007) Design, analysis and fabrication of Microflexural not gate. ASME 2007 International Mechanical Engineering Congress and Exposition, Seattle, Washington, vol. 11. pp 501–508

Peterson K (1982) Silicon as a mechanical material. Proc IEEE 70(5):420–457

Yang X, Tai YC, Ho CM (1997) Micro bellow actuators. Technical digest, 1997 international conference on solid-state sensors and actuators (Transducers '97), vol 1, Chicago, pp 45–48

Stability Analysis of Semi-kinematic Mountings used in Modular Reconfigurable Micro Factory Testbed

Mounika Katragadda, Aneissha Chebolu, and Nagahanumaiah

Abstract In this paper three groove semi kinematic mounting has been designed and analyzed its the stability in micro milling operational conditions, employed in a specific modular reconfigurable micro factory test bed developed at authors group. While various kinds of kinematic couplings, which are popular for their high repeatability and interchangeability, are considered for the modular machine design, Gothic arch and Maxwell coupling designs were compared. In this evaluation process, first the maximum preload, which can resist the induced contact stresses, is determined. Then the static stiffness matrices are evaluated for each case and compared as a measure of their relative stability. Maxwell coupling which was found to be best was adopted for the machine stage and experiments were performed to know the coupling stability with varying load.

Keywords Semi Kinematic mounts • Modular machines • Micro factory • Stability analysis

1 Introduction

In order to meet the growing demand for the development of miniaturized devices with microscopic and sub-microscopic patterns and structures on multi materials, the micro manufacturing technologies have witnessed several manufacturing capabilities. Given that micro-nano systems engineering is still an emerging field, the current requirement of manufacturing systems can be characterized as a multi-functional, multi-scale flexible manufacturing system for small order production. Therefore the concept of reconfigurable manufacturing systems (RMS) evolved by combining the features of dedicated and flexible manufacturing systems is expected to be a future technology, where a family of products can be rapidly produced. However, success of RMS in micro-nano scale manufacturing would impose

M. Katragadda • A. Chebolu • Nagahanumaiah (✉)
Micro Systems Technology Laboratories, CSIR-Central Mechanical Engineering Research Institute, Durgapur, India
e-mail: mounika.118@gmail.com; aneissha.mech@gmail.com; naga@cmeri.res.in

E.-C. Lovasz et al. (eds.), *Microactuators and Micromechanisms*, Mechanisms and Machine Science 30, DOI 10.1007/978-3-319-15862-4_14

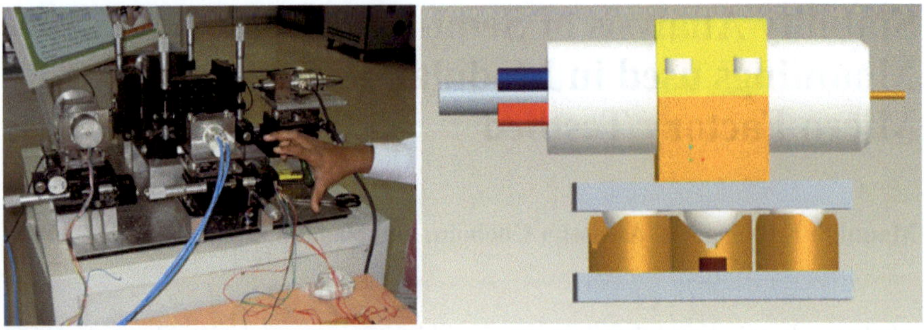

Fig. 1 Micro factory testbed and semi kinematic mounts used for modularity

several challenges with respect to modular micro machines design and technical maturity of the manufacturing processes (process repeatability).

Authors group has recently developed a micro-factory test bed to perform four operations: μ-milling, μ-turning, μ-EDM and μ-laser ablation. This test bed is shown in Fig. 1 and incorporates design modularity, multi-operational capabilities and desktop manufacturing system with 600 mm × 600 mm footprint capable of machining features as small as 100 μm over multi materials. These machining units may be reconfigurable for a family of products. To satisfy the constraint of modularity and precision positioning on its ramp-up, kinematic couplings are chosen. In this paper, the stability of three groove kinematic mounts used for machining stages has been investigated for horizontal micro milling configuration.

Good kinematic design seeks to utilize only the minimum number of DOF to constrain the motion of a body. The closure force for the application of kinematic design can be gravity, or spring loading or calmping, but must be applied through the kinematic contacts. However, there are certain disadvantages associated – load carrying limitation, stiffness may be too low for application and low damping. These can be overcome by semi-kinematic approach. The direct way is to replace all point contact with a small area. Doing so decreases the contact stress, but increases the stiffness and load carrying capacity.

In literature, two basic kinematic couplings are widely studied. Maxwell coupling, which consists of three radial Vee-grooves on one plate and three balls on the other plate and Kelvin coupling, which has its three balls rested on three geometries, viz., trihedral nest, a Vee-groove and a flat plate. But Kelvin coupling is a non-symmetric coupling and is not suitable for high speed operations. Maxwell configuration is preferred when the orientation of the v-grooves is to be decided. This can be done with the help of Maxwell's criterion (Slocum 1992). For balanced stiffness in all directions, the contact force vectors should intersect the plane of coupling action at an angle of 45°. Besides, when there is a case of external loading, stability is ensured by applying a preload. Here, the external loading is from the cutting forces generated by micro milling operations. Apart from these basic configurations, Slocum proposed a Gothic arch design for the vee-grooves (Slocum 1988), instead of their flat sides. When this configuration is used, the effect of contamination at the interface due to repeated loading would be less. Although, Maxwell couplings are preferred for their easy manufacturability, stability of both Maxwell coupling and Gothic arch design have been investigated in this work.

2 Determination of Preload

Forces acting on the contact points and total system (reaction, friction, external forces) are listed below. The notations assumed are shown in the Fig. 2.

In the equations 'P' represents the reaction force, 'Fs' represents frictional force along the side of the groove, 'Ft' represents frictional force along the axis of the groove and 'R' is the magnitude of the reaction force. Also, x is the groove angle, y is the angle involved in coupling triangle and Wt is the weight of the structure. Coefficient of friction is denoted with 'u' [Here u is considered as 0.3]. The maximum cutting forces that can be generated in the micro milling process are 10N (perpendicular to feed direction, in horizontal plane), 7N in feed direction and 4N (perpendicular to feed direction, in vertical plane). The equilibrium equations were solved for the reaction and frictional forces using MATLAB.

$$\begin{bmatrix} P_{1a} \\ P_{1b} \\ P_{2a} \\ P_{2b} \\ P_{3a} \\ P_{3b} \end{bmatrix} = \begin{bmatrix} R_{1a}\cos x \sin y & R_{1a}\sin x & R_{1a}\cos x \cos y \\ -R_{1b}\cos x \sin y & R_{1b}\sin x & -R_{1a}\cos x \cos y \\ R_{2b}\cos x \sin y & R_{2a}\sin x & -R_{2a}\cos x \cos y \\ -R_{2b}\cos x \sin y & R_{2b}\sin x & -R_{2b}\cos x \cos y \\ R_{3a}\cos x & R_{3a}\sin x & 0 \\ -R_{3b}\cos x & R_{3b}\sin x & 0 \end{bmatrix} \tag{1}$$

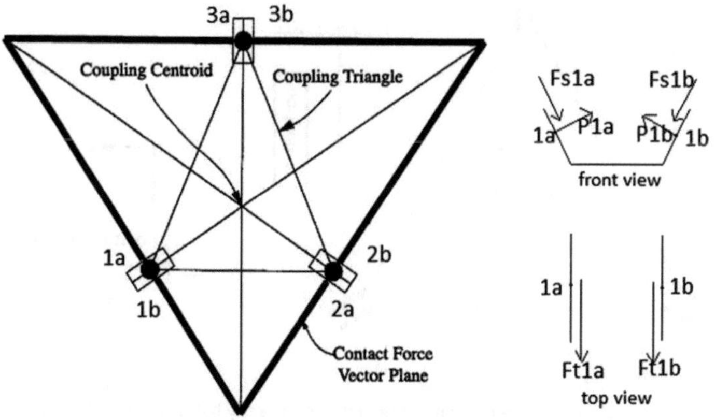

Fig. 2 Planar kinematic coupling layout

$$\begin{bmatrix} Fs_{1a} \\ Fs_{1b} \\ Fs_{2a} \\ Fs_{2b} \\ Fs_{3a} \\ Fs_{3b} \end{bmatrix} = \begin{bmatrix} -u \times R_{1a}\text{sinx siny} & u \times R_{1a}\text{cosx} & -u \times R_{1a}\text{sinx cosy} \\ u \times R_{1b}\text{sinx siny} & u \times R_{1b}\text{cosx} & u \times R_{1a}\text{ sinx cosy} \\ -u \times R_{2a}\text{sinx siny} & u \times R_{2a}\text{cosx} & -u \times R_{2a}\text{sinx cosy} \\ u \times R_{2b}\text{sinx siny} & u \times R_{2b}\text{cosx} & u \times R_{2b}\text{ sinx cosy} \\ -u \times R_{3a}\text{sinx} & u \times R_{3a}\text{cosx} & 0 \\ u \times R_{3b}\text{sinx} & u \times R_{3b}\text{cosx} & 0 \end{bmatrix} \quad (2)$$

$$\begin{bmatrix} Ft_{1a} \\ Ft_{1b} \\ Ft_{2a} \\ Ft_{2b} \\ Ft_{3a} \\ Ft_{3b} \end{bmatrix} = \begin{bmatrix} -u \times R_{1a}\text{cosy} & 0 & u \times R_{1a}\text{siny} \\ -u \times R_{1b}\text{cosy} & 0 & u \times R_{1b}\text{ siny} \\ -u \times R_{2a}\text{cosy} & 0 & u \times R_{2a}\text{siny} \\ -u \times R_{2b}\text{cosy} & 0 & u \times R_{2b}\text{siny} \\ 0 & 0 & u \times R_{3a} \\ 0 & 0 & u \times R_{3b} \end{bmatrix} \quad (3)$$

$W = [0 \quad w_t \quad 0]$ and the external forces are $F_{ext} = [F_x \quad F_y \quad F_z]$.

The maximum stability condition is achieved when the cutting forces are in a direction as shown in the Fig. 3. For this condition, the preload required for the system to be stable is 15N, i.e., 1.5 kg, in addition to the self weight. A magnetic preload is always preferred. But instead, a maximum preload which doesn't affect the materials can be applied. As the cutting forces are dynamic in nature, this extra preload can compensate the induced instability. The reaction force values at the six contact points were; $R_{1a} = 8.6047N$, $R_{1b} = 5.2472N$, $R_{2a} = 9.8148N$, $R_{2b} = 2.5391N$, $R_{3a} = 15.7509N$, $R_{3b} = 0.0781N$.

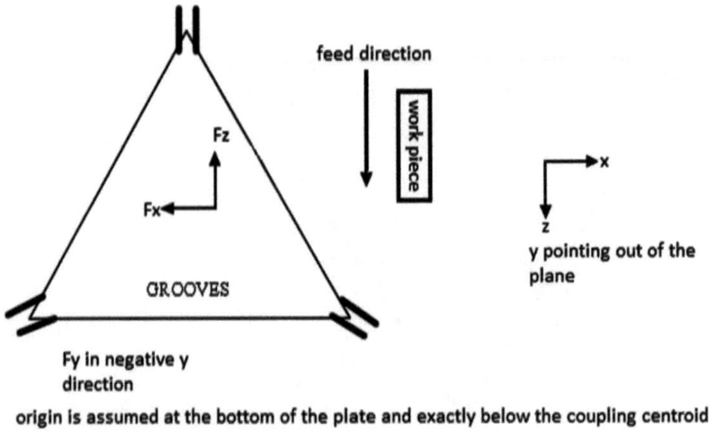

Fig. 3 Maximum stability condition

3 Stiffness and Stress Evaluation

Hertz theory (Johnson 2003) generally refers to a point contact problem and can be used to know the stresses at every contact point and stiffness in different directions. The following assumptions are made in determining the solutions of Hertzian contact problems:

- The strains are small and within the elastic limit,
- Each body can be considered an elastic half-space, i.e., the area of contact is much smaller than the characteristic radius of the body,
- The surfaces are continuous and non-conforming, and
- The surfaces are frictionless.

Here, as the surfaces are also affected by friction, a deformation in tangential direction is also considered in addition to normal deformation.

In the case of flat sided vee groove R = Rb, and for gothic arch, R = $Rb(1+\eta)/(1+2\eta)$, where $\eta = 0.2$ (Slocum 1988).

Equivalent Youngs modulus, Ee (or) E* is,

$$E^* = \left(\frac{1-{\nu_1}^2}{E_1} + \frac{1-{\nu_2}^2}{E_1}\right)^{-1} \tag{4}$$

Equivalent area of contact, $a = \left(\frac{3 \cdot F_N \cdot R}{4 \cdot E^*}\right)^{1/3}$; F_N = contact force between 2 bodies

Deflection in normal direction is,

$$\delta_z = \frac{a^2}{R} = \left(\frac{9 \cdot {F_N}^2}{16 \cdot R \cdot E^{*2}}\right)^{1/3} \tag{5}$$

Contact pressure,

$$q = \left(\frac{1}{\pi}\right)\left(\frac{1}{\text{Re}}\right)^{\frac{1}{3}}\left(\frac{3Ee^2}{2}\right)^{(1/3)} = (\text{aEe})/\,(\pi\,\text{Re}) \tag{6}$$

For tangential forces (Van Veggel and Nijmeijer 2008): The relative tangential displacement of the contacting bodies in x-direction (tangential direction) due to the tangential load is given by:

$$\delta_x = \frac{3 \cdot \mu \cdot F_N}{16 \cdot a} \cdot \left(\frac{2-\nu_1}{G_1} + \frac{2-\nu_2}{G_2}\right) \cdot \left\{1 - \left(1 - \frac{F_T}{\mu \cdot F_N}\right)^{2/3}\right\} \tag{7}$$

Where G_1 the shear modulus of the sphere [N/m^2], and G_2 the shear modulus of the flat [N/m^2]. But as slip has already occurred, $F_T = \mu$ FN.

Stiffness K = F/x, the force-deformation equation with overall stiffness can be represented as:

$$\begin{bmatrix} f \\ -- \\ m \end{bmatrix} = \begin{bmatrix} \mathbf{Kf}/\delta & | & \mathbf{Kf}/\theta \\ -- & | & -- \\ \mathbf{Km}/\delta & | & \mathbf{Km}/\theta \end{bmatrix} \cdot \begin{bmatrix} d\delta \\ -- \\ d\theta \end{bmatrix} \tag{8}$$

A rotation-translation transformation matrix is used to convert the local stiffness matrices into global coordinate system (Hale and Slocum 2000).

$Kx = F$; $K(T^*x) = (T^*F)$ and $K_{global} = \left(T^T K^* T\right)$, if K is the local stiffness matrix.

The evaluated stiffness matrix of the structure in Maxwell configuration is,

$$K = \begin{bmatrix} 1.8426 & 1.9482 & 1.8401 & -0.0004 & -0.0004 & -0.0004 \\ 1.9482 & 2.0660 & 1.9571 & 0.0011 & 0.0011 & 0.0011 \\ 1.8401 & 1.9571 & 1.8647 & 0.0007 & 0.0008 & 0.0008 \\ -0.0004 & 0.0011 & 0.0007 & -0.0004 & -0.0006 & -0.0008 \\ -0.0004 & 0.0011 & 0.0008 & -0.0006 & -0.0007 & -0.0010 \\ -0.0004 & 0.0011 & 0.0008 & -0.0008 & -0.0010 & -0.0011 \end{bmatrix} \times 10^8$$

The evaluated stiffness matrix of the structure in Gothic arch configuration is,

$$K = \begin{bmatrix} 1.7503 & 1.8506 & 1.7479 & -0.0003 & -0.0004 & -0.0003 \\ 1.8506 & 1.9625 & 1.8591 & 0.0010 & 0.0011 & 0.0010 \\ 1.7479 & 1.8591 & 1.7713 & 0.0007 & 0.0007 & 0.0007 \\ -0.0003 & 0.0010 & 0.0007 & -0.0004 & -0.0006 & -0.0007 \\ -0.0004 & 0.0011 & 0.0007 & -0.0006 & -0.0007 & -0.0009 \\ -0.0003 & 0.0010 & 0.0007 & -0.0007 & -0.0009 & -0.0011 \end{bmatrix} \times 10^8$$

It can be seen that stiffness for Gothic arch is less than the Maxwell configuration. Further, manufacturing and cleaning the Gothic arch is time consuming and difficult compared to flat Vee-groove, therefore Maxwell configuration couplings it is suggested to opt for the flat sided vee groove.

4 Stability Analysis by Varying Load

Experiments were conducted to know the behavior of the coupling when load is gradually varied (Fig. 4). As it is practically difficult to measure the displacement of top plate that carries high speed spindle in-situ in milling operation, instability of the mounts under varying load was done offline. Load is applied in two directions; negative X-direction and negative Z-direction. To take measurements a sharp tip is mounted vertically at the center of the coupling plate, at a height of 519 mm from

Fig. 4 Experimental set up showing lost point contact

the base of the coupling. Each time the load is varied, the displacement at the tip were measured. This was done by setting up an identical tip against it. When there was a deflection in lower tip the upper tip was adjusted such that it always coincided with the lower one. A coordinate measuring instrument was attached to the upper tip to help in its movement. As there was no availability of load cell, a string was tied to the coupling on one side and to a bottle on other side. Load was varied by varying the amount of water in the bottle.

Displacement in X direction has been found to be higher with varying load. Displacement in Y direction is almost always less than the deflection in the other two directions. As the load was not gradually varied, but always removed and applied again, uneven fluctuations in the displacement might have occurred. The displacement band is compact in the case of varying load in X direction compared to the variation of load in Z direction. This shows that the precision machining is still possible even if forces in X direction vary. It is also observed that one of the point contacts (Fig. 4) is lost when a load of 19N is applied in –ve X-direction and a load of 19.46N in Z-direction. In Fig. 4 the location of contact loss is shown. But theoretical calculations show that, to lose contact at 2a, loads must be 5N and 15N in – ve X and Z directions respectively. This is believed to be because of the string elasticity and because of the uneven friction present at the six contacts. So, this should be avoided.

5 Conclusions

It is appreciated that the deployment of micro-nano systems/devices for rapidly growing applications such as optical systems, solar cells and biomedical devices, demand for modular micro machines that could be reconfigurable for family of parts. In this paper, stability of semi-kinematic mounts under the micro machining forces has been investigated. It has been found that Maxwell coupling configuration is more stable than the Gothic design for this specific micro factory system design. While the requirement of pre-loading of 15N is taken care by the self weight of high speed spindle and its mounting elements, experiments were performed by varying loads in other two axes. These mountings were stable, the deflection measured in Cartesian coordinates indicate that change in force in X-direction (feed force in micro milling) has significant effect than that of normal force (along Z-axis, i.e. tool axis). The designed Maxwell semi kinematic mounts could withstand forces upto $F_x = 19N$ and $F_z = 19.4N$ that ensures precision machining at normal micro milling conditions of $F_x = 10N$; $F_z = 7N$ and $F_y = 4N$. It has been demonstrated that semi-kinematic mounts which are otherwise widely used in optical equipments can be adopted in the design of modular machines used for micro manufacturing. Further, these mounts allows quick ramp-up that make reconfigurable micro manufacturing systems a reality for effective and quick fabrication of a family of micro-nano scale products in batch scale.

References

Hale L, Slocum AH (2000) Optimal design techniques for kinematic couplings. Precis Eng 24 (2):114–127

Johnson KL (2003) Contact mechanics. Cambridge University Press, Cambridge/New York

Slocum AH (1988) Kinematic couplings for precision fixturing – Part I: formulation of design parameters. Precis Eng 10(2):85–91

Slocum AH (1992) Design of three groove kinematic couplings. Precis Eng 14(2):67–76

Van Veggel AA, Nijmeijer H (2008) Stable mounting of beam splitters for an interferometer. Precis Eng 33(1):7–17

Multibody System Simulation of Hysteresis Effect by Micro Textile-Reinforced Compliant Mechanisms with Piezo-electric Actuators

D. Mărgineanu, E.-C. Lovasz, and N. Modler

Abstract The function integration in textile reinforced compliant structures requires integrated actuation besides mobility and sensors. Especially in micro-mechanisms, the piezo-actors are very useful due to their compactness, high forces and electrical control. However, the actuation hysteresis may be a problem in designing such structures aiming specific movements. Some procedures for multibody simulation taking into account the piezo-electric hysteresis effect are presented in this paper.

Keywords Piezo-actuators • Hysteresis • Micro-mechanisms • Multi body system simulation

1 Introduction

Compliant mechanisms (CM) consist of elements with higher rigidity and compliant elements, with higher elasticity according to their functional purpose (Howell 2001). Also, classic rotational joints in hybrid structures or elastic joints in monolithic structures may be used. CM can fulfill desired functions with minimum number of elements and joints. Textile reinforced structures are especially suited for building CM, as they may have variable rigidity.

By integrating sensors and actuators in the composite structure, active textile reinforced compliant mechanisms (A-TCM) are obtained, further reducing the product complexity and costs. Piezo-electric actuators, with their compactness and high actuating forces, have certain advantages by using them in ATCM, as allowing direct electrical control (Elspass and Flemming 1998; Hufenbach et al. 2006b).

D. Mărgineanu (✉) • E.-C. Lovasz
Department of Mechatronics, University Politehnica Timisoara, Timisoara, Romania
e-mail: dan.margineanu@upt.ro; erwin.lovasz@upt.ro

N. Modler
Institut für Leichtbau und Kunststofftechnik, Technical University Dresden, Dresden, Germany
e-mail: niels.modler@tu-dresden.de

E.-C. Lovasz et al. (eds.), *Microactuators and Micromechanisms*, Mechanisms and Machine Science 30, DOI 10.1007/978-3-319-15862-4_15

The A-CM transmission function and the trajectories, depending both on elements geometry and on the loads, and are strongly influenced by the elastic and inertial properties of the materials used (Hufenbach et al. 2006a; Modler 2008; Midha et al. 2000), thus the desired accurate motion can be obtained only after tests, re-design and adjustments, implying high costs and consuming precious time (Modler et al. 2009).

The increasing number of compliant mechanisms industrial applications led to researches on design approaches for developing compliant mechanisms for required motion and force-deflection characteristics (Hufenbach and Gude 2002; Hufenbach et al. 2007). The Pseudo-Rigid-Body Model (PRBM) allows compliant mechanisms to be modeled and analyzed as rigid-body mechanisms and significantly reduces the analysis complexity.

Parallel mechanisms, a bicycle derailleur and parallel-motion bicycle brakes are presented in (Mattson et al. 2004). Compliance comparison between multi-spring model and finite-element analysis for multidimensional acceleration sensors are done in (Gao and Zhang 2010). Analysis and design of an under-actuated compliant variable stroke mechanism by employing PRBM is studied in (Tanık and Söylemez 2010).

A-CM with various functioning principles are also extending their application range. On-chip sensing of bi-stable mechanism state using the piezo-resistive properties of poly-silicon (Anderson et al. 2006), is made by detection of changes in position by changes in resistance across the mechanism.

A piezoelectric driven compliant-based micro-gripper mechanism for manipulation of micro objects, shown by (Nashrul et al. 2009) address the nonlinearities associated with the application of conventional rigid hinges.

By A-TCM, analytical calculations can only give rough approaches because the calculus models include partial differential equations systems, the initial and limit condition depending on the parts geometry, yet to be determined. Or, the calculus model is simplified thoroughly, thus the results are not accurate enough.

A simulation procedure should be carried out between the analytic dimensioning and the FEM modeling step.

This simulation should observe some basic requirements:

- all the initial simulation parameters should be defined according to the dimensions, materials and loads of the simulated system;
- the simulation results should estimate with an accepted limit error the behavior of the real system;
- the simulation parameters should be easily changed, also according to optimization procedures;

The simulation procedure should be fast, reliable and repeatable.

2 Modeling A-TCM for Simulation

Piezoelectric energy converters provide high forces, response times of a few milliseconds and a positioning accuracy of a few nanometers, but only at more than one to two thousand Volt. In Eq. (1), by inverse piezoelectric effect used for actuation, the mechanical strain S depends on mechanical load T and electric field strength E. If the tensor of mechanical parameters on the surface of a crystal is considered, the linear equation in vector form is, in Voigt's notation is:

$$\{S\} = \left[s^E\right] \cdot \{T\} + \left[d^t\right] \cdot \{E\} \tag{1}$$

where:

$[d^t]$ is the matrix for the reverse piezoelectric effect,
ε^T – permitivity Matrix for $T = 0$,
s^E – compliance matrix for $E = 0$.

So, an accurate simulation model for an A-TCM should ensure both an adequate compliance and an actuation depending both on the input signal and on the effort in the actuator.

As the stroke of a piezo-actuator is very small, bending actuators with two layers: one passive and one active (see Fig. 1) are used. In this paper, an actuator with carbon fiber composite (CFC) as passive base layer and micro fiber composite (MFC) as actuator layer is considered.

The compliant RBS simulation model proposed for the actuator is composed of some simulation cells linked base-to-end to one another.

In order to better describe the piezoelectric actuation which adds the actuating and elastic deformations, the model presented in Fig. 2 is introduced. The 3 DOF simulation cell is a 6-bar mechanism with 4 revolute joints and 2 prismatic joints.

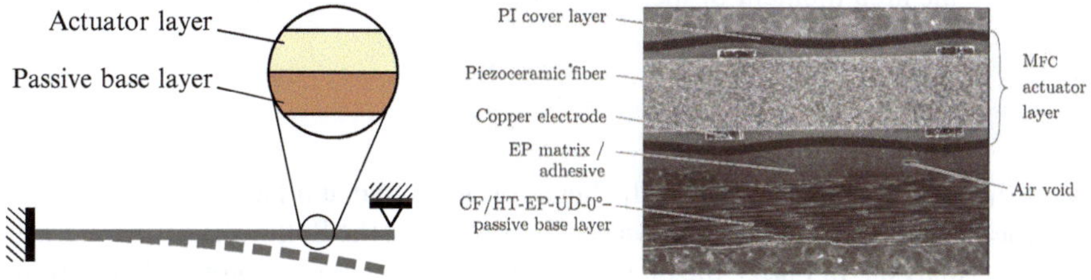

Fig. 1 The active element layers

Fig. 2 Active element cell kinematical model with 3 DOF and 2 prismatic joints simulating the piezo-actuating effect

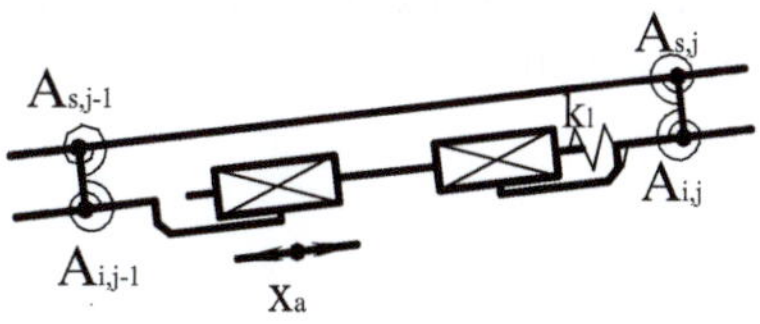

The active equivalent layer $A_{i,j}A_{i,j+1}$ is composed of 3 elements, linked by 2 prismatic joints. The actuation x_a in the active joint obtained by applying the voltage U will be:

$$x_a = \Lambda \cdot L_j \tag{2}$$

with the actuating deformation Λ given in Eq. (3) with respect to input voltage U:

$$\Lambda = d_{33} \cdot E = d_{33} \cdot \frac{U}{t_{IDE}} \tag{3}$$

with d_{33}- piezoelectric constant, and t_{IDE} – distance between electrodes.

The elastic deformation of the actuator piezo-ceramic layer is simulated by the helical compression spring with the rigidity k_l:

$$k_l = E_2 \cdot \frac{b_2 \cdot h_a}{L_j} \tag{4}$$

The active layer is joined to the interface element in the rotational joints $A_{i,j}$ and $A_{i,j+1}$. In order to simulate the bending compliance of the two layer actuator, torsion springs are inserted in the upper and lower $A_{s,i;j}$ rotational joints. As two springs are connected on each side of the interface to ensure equal angles with the current an the previous layer element, the rigidity of these springs is calculated as the double of the equivalent bending rigidity of the CFK and MFC layers:

$$k_{\theta s,i} = \frac{M_j}{\theta_{s,i;j}} = E_{1,2} \cdot \frac{I_{z_{1,2}}}{L_j}, \tag{5}$$

where

- $I_{z1,2}$ – the area moment of inertia of each of the two layers cross-section:

$$I_{z_{1,2}} = \frac{b_{1,2} \cdot {h_{1,2}}^3}{12}, \tag{6}$$

with $b_{1,2}$, the width and $h_{1,2}$, the thickness of the active and passive layers.

The modeling and simulation was performed using the Solidworks software with Motion Analysis. In Fig. 3, the model of the bending piezo-actuator built according to this procedure is shown.

The actuation x_a of the linear motor in the simulation cell as a time-dependent function to be used in simulations will be determined in the following, taking into consideration the hysteresis effect.

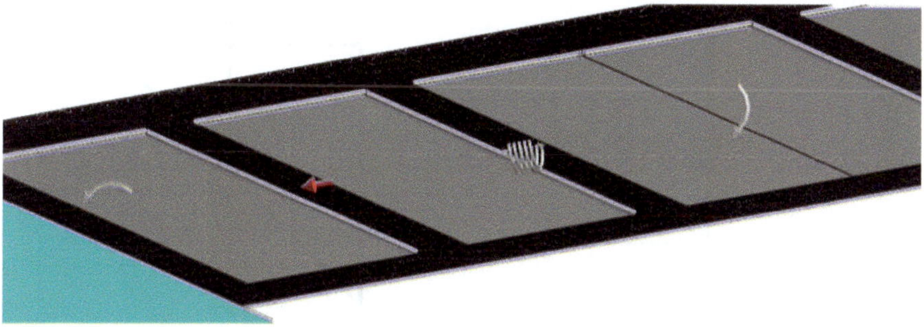

Fig. 3 The 3d model of the simulation cell for the bending piezo-actuator

3 Continuous Preisach Model Simulation of Piezo-electric Hysteresis

In order to simulate the hysteresis of the piezo-actuator, a procedure to convert the input voltage signal in the displacement output with hysteresis is needed.

The continuous Preisach model for hysteresis describes the output function as:

$$y(t) = \Gamma \cdot x(t) = \iint_{\beta \geq \alpha} \mu(\alpha, \beta)\ R_{\alpha,\beta}\ x(t)\ d\alpha\ d\beta. \tag{7}$$

where y(t) is the output signal, x(t) is the input signal, μ(α,β) is the density function and R(α,β) is the hysteron (see Figs. 4 and 5).

For equal density continuous Preisach operator, the integral reduces to the aria of a triangle in the Preisach plane, i.e. a 2nd degree input/output dependency.

So, for a continuous variation of the input voltage from the minimum to the maximum value and back, the output deformation may be approximated with Eq. (8), as the voltage rises from U_{min} to U_{max} (see also Fig. 5),

$$x_1(U) = x_{min} + (x_{max} - x_{min}) \cdot \left(\frac{U - U_{min}}{U_{max} - U_{min}}\right)^{r_1} \tag{8}$$

and, for the voltage falling from U_{max} to U_{min}, with Eq. (9)

$$x_2(U) = x_{max} - (x_{max} - x_{min}) \cdot \left(\frac{U_{max} - U}{U_{max} - U_{min}}\right)^{r_2} \tag{9}$$

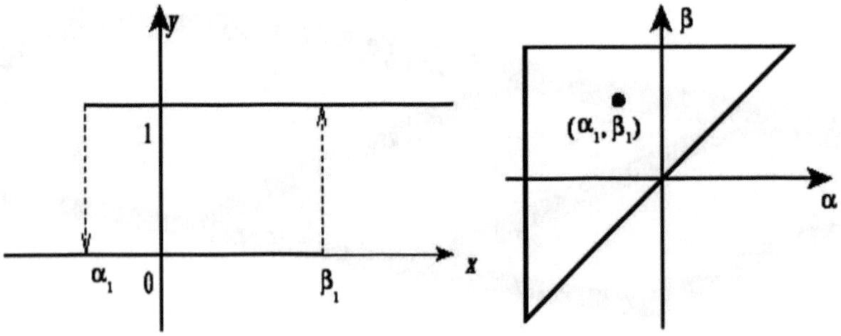

Fig. 4 Non-ideal relay and its corresponding point on the Preisach plane

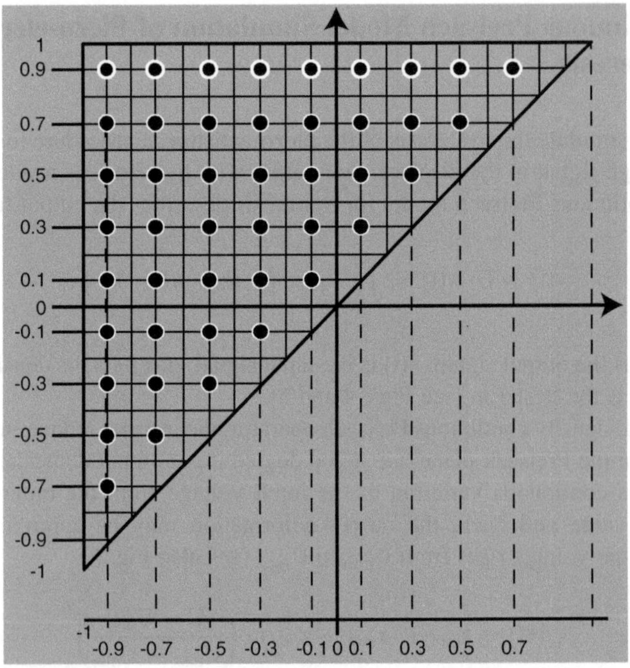

Fig. 5 Geometrical interpretation of Preisach nonlinearity with standard discrete support

The exponents $r_{1,2}$, calculated with Eq. (10), are used to adjust the approximation curves to fit better to the experimental ones by passing through the third precision point ($U_{m1,2}$; $x_{m1,2}$):

$$r_1 = \frac{\ln \frac{x_{m1} - x_{min}}{x_{max} - x_{min}}}{\ln \frac{U_{m1} - U_{min}}{U_{max} - U_{min}}}; \ r_2 = \frac{\ln \frac{x_{max} - x_{m2}}{x_{max} - x_{min}}}{\ln \frac{U_{max} - U_{m2}}{U_{max} - U_{min}}} \tag{10}$$

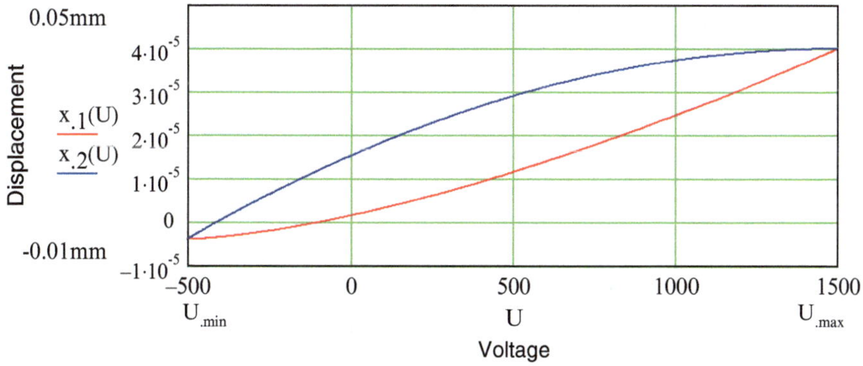

Fig. 6 The hysteresis cycle approximated with the continuous Preisach model

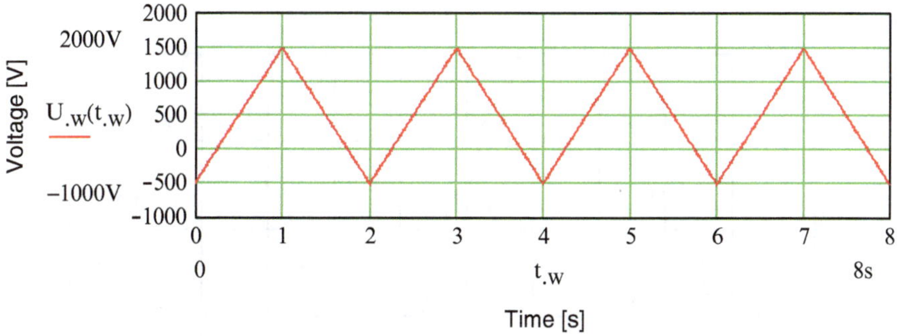

Fig. 7 Saw tooth input signal

For a simulating cell 29 mm long, $U_{max} = 1{,}500$ V max voltage, $d_{33} = 4.6\ 10^2$ pC/N piezoelectric constant and $t_{IDE} = 0.5$ mm distance between electrodes, the total actuator stroke calculated with Eqs. (2) and (3) is 40 μm. The negative displacement at $U_{min} = -500$ V is – 4 μm, i.e. 10 % of the maximal displacement. Approximation curves with $r_1 = 1.5$ and $r_2 = 2$ calculated for $U_m = 500$ V, $x_{m1} = 12$ μm and $x_{m2} = 28$ μm are presented in Fig. 6.

For a saw tooth voltage input signal in Eq. (11) and Fig. 7, the output piezoelectric displacement cycle of the active layer with hysteresis effect with Eqs. (11) and (12) will be as in Fig. 8.

$$U(t) = U_{med} + U_{amp} \cdot \left(\frac{4t}{T} - 1 - 2\mathrm{int}\left(\frac{2t}{T}\right)\right) \cdot (-1)^{\mathrm{int}\left(\frac{2t}{T}\right)} \tag{11}$$

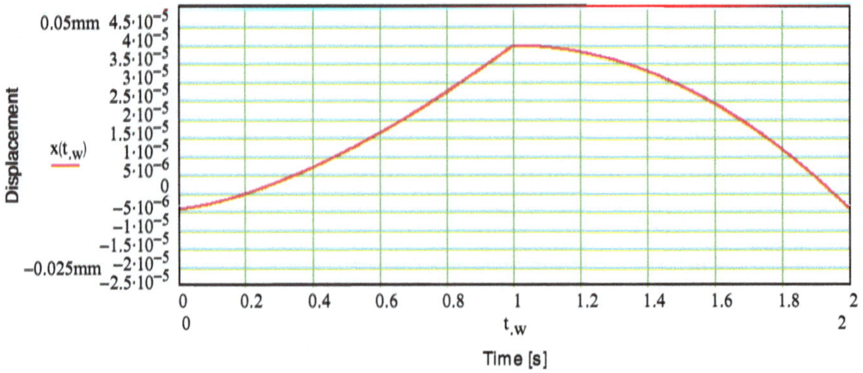

Fig. 8 Approximated piezo-electric displacement output signal for saw tooth input signal

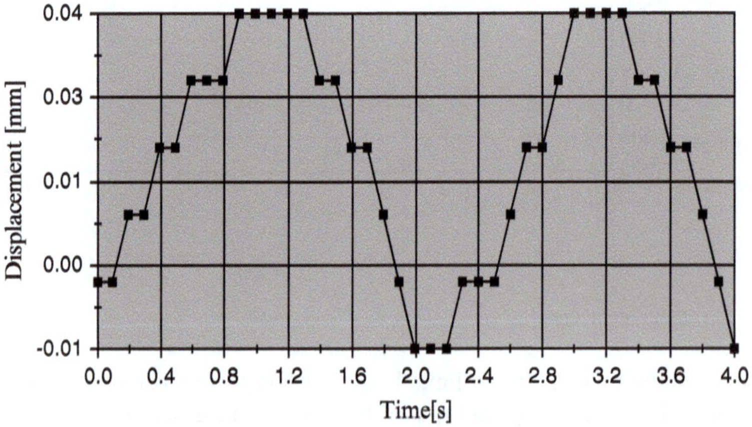

Fig. 9 Displacement versus time points for approximated piezo-electric displacement output signal for saw tooth input signal

For simulations, a table of displacements versus time is generated to be used by controlling the linear motor in the simulation cell by interpolation, as in Fig. 9. The command window for the linear motor of the simulating cell is presented in Fig. 10.

The simulation model for the bending actuator built with the proposed simulation cells is presented in Fig. 11, and the simulation results are shown in Fig. 12.

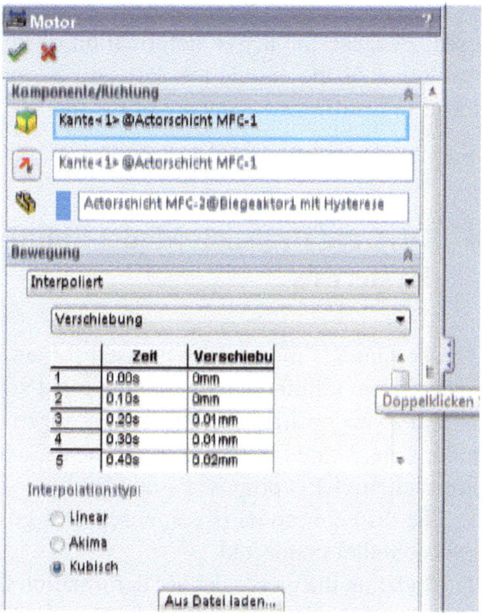

Fig. 10 The command window for the linear motor of the simulating cell

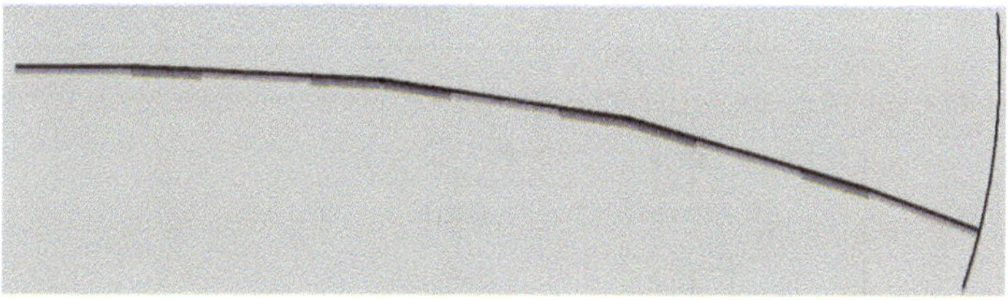

Fig. 11 Bending actuator simulation model

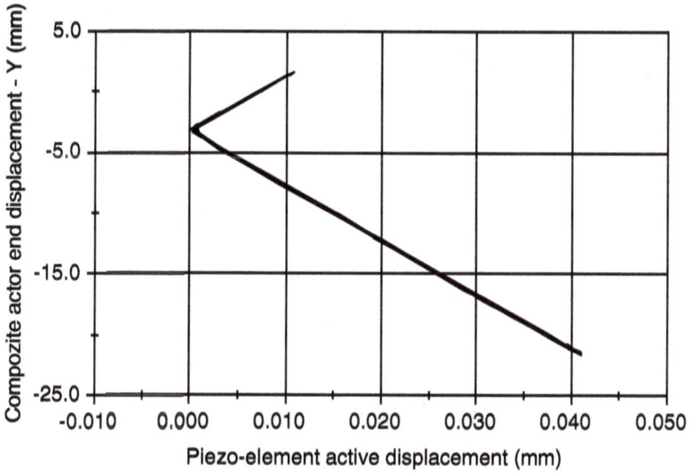

Fig. 12 Simulation results for the bending actuator with hysteresis

The simulation is more accurate as more precision points on the action table are used. Because the active deformation of linear piezo actuators is very small, the resolution of the simulation software is important. Resolutions of 0.01 mm may result in insufficient accuracy of the model.

4 Discrete Preisach Model Simulation of Piezo-electric Hysteresis

The continuous model may be used for simulating A-TCM behavior in complete cycles, from minimum to maximum input voltage. As to simulate mechanisms with intermediate positions controlled by piezo actuators, where the hysteresis gives more complicated output displacements, a simulation procedure using a discrete Preisach model is proposed (see Fig. 13).

The discrete model is composed of a group of non ideal relays and adjustable gains parallel connected.

Modeling the piezo-electric deformation Output signal with hysteresis is done by adjusting the switch on and the switch off points of the Relays 1–6 and the Gains 1–6.

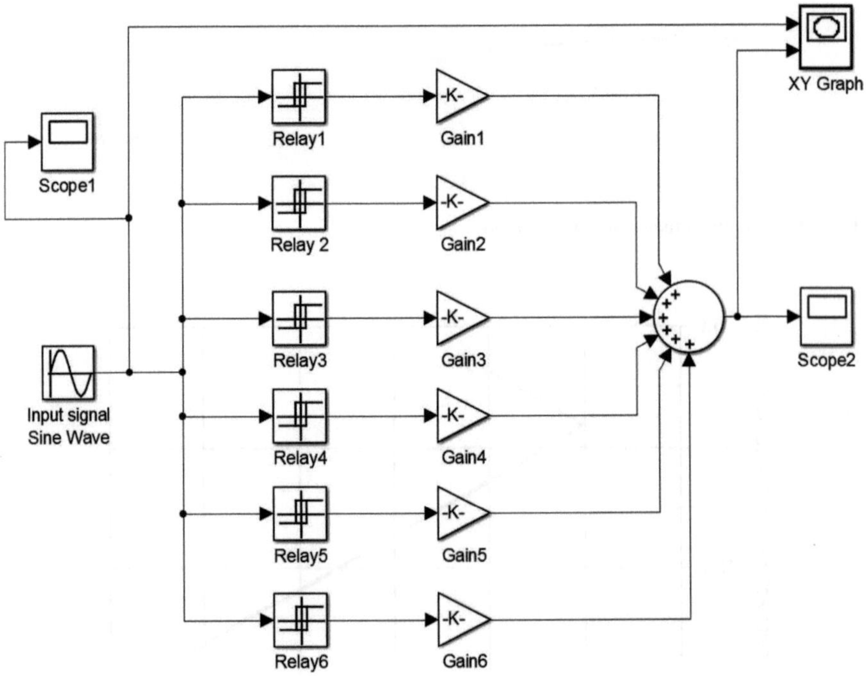

Fig. 13 Matlab Simulink model for discrete Preisach piezo-electric hysteresis simulation

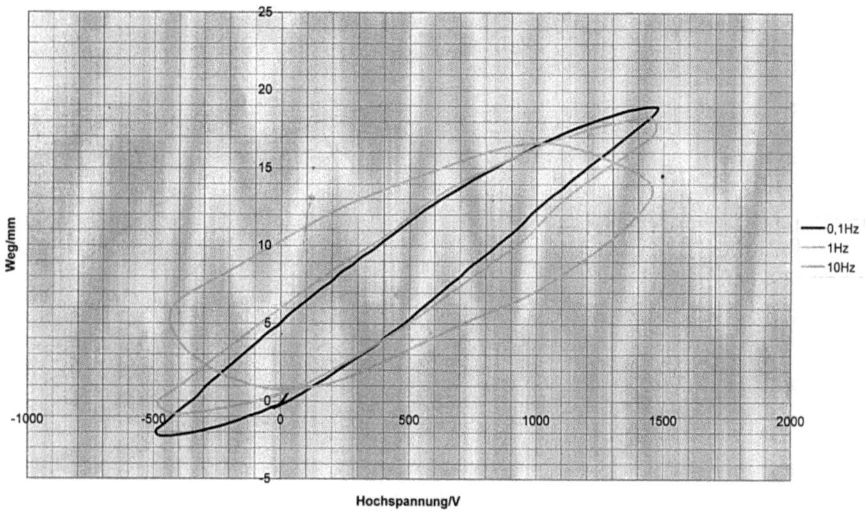

Fig. 14 Experimentally determined piezo-electric hysteresis

There strategy for determining the switch on/switch off points and gains is to have equal gains Δy for the n relays.

$$\Delta y = \frac{y_{max} - y_{min}}{n} \tag{12}$$

The switch on/switch off points are determined from the experimental hysteresis curve (Fig. 14) as the abscissa of the intersection points between the curve and horizontal lines at $y_i = i\ \Delta y$. The left point abscissa is the switch off point and the right point abscissa is switch on point of the relay.

This procedure insures a good approximation for the steep parts of the hysteresis curve, but lower accuracy for zones with lower slope.

In order to optimize the approximation without increasing the number of switches, a procedure with variable gains according to the local slope of the curve is to be developed.

The hysteresis cycle approximated with the discrete Preisach model is presented in Fig. 15.

The approximated input/output dependency is used to control the linear motor in the simulation cell. This model may be used in case of an input signal variable within the input range, with ups and downs not necessarily to limits.

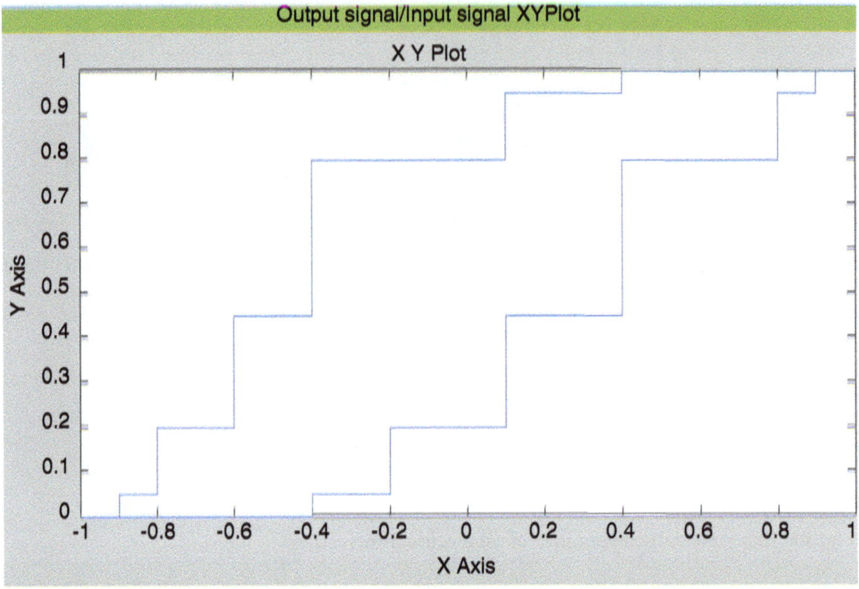

Fig. 15 X-Y plot of an input/output dependency generated with the discrete Preisach model for simulation of piezo-electric hysteresis

5 Conclusions

The procedure for simulation of the piezo-actuators with hysteresis presented in this paper allows the designer to approach the behavior of A-TCM in the mechanism synthesis stage.

The hysteresis effect may be important in function or path generating mechanisms and is not to be neglected.

The continuous Preisach model for hysteresis is useful for stable input voltage signals oscillating stably between minimal and maximal values.

For variable input signals, the discrete Preisach model may provide a simulation tool able to predict the mechanism's behavior in the synthesis stage of the development procedure.

Acknowledgment The authors would like to express their gratitude towards the Deutsche Forschungsgemeinschaft (DFG), which supports this research within the scope of the subproject D2 of the Collaborative Research Centre SFB 639 "Textile-Reinforced Composite Components in Function-Integrating Multi-Material Design for Complex Lightweight Applications".

References

Anderson JK, Howell LL, Wittwer JW, McLain TW (2006) Piezoresistive sensing of bistable micro mechanism state. J Micromech Microeng 16:943–950. doi:10.1088/0960-1317/16/5/010 30

Elspass WJ, Flemming M (1998) Aktive Funktionsbauweisen: Eine Einführung in die Struktronik. Springer, Berlin/Heidelberg/New York

Gao Z, Zhang D (2010) Design, analysis and fabrication of a multidimensional acceleration sensor based on fully decoupled compliant parallel mechanism. Sensors Actuators A 163:418–427

Howell LL (2001) Compliant mechanisms. Wiley, New York

Hufenbach W, Gude M (2002) Analysis and optimisation of multi-stable composites under residual stresses. Compos Struct 55:319–327

Hufenbach W, Modler K-H, Täger O, Modler N, Lovasz E-C (2006a) Design and manufacturing of smart textile compliant hinges. Acta Tech Napoc 2:831–836

Hufenbach W, Modler K-H, Täger O, Modler N, Renner O (2006b) Contribution to the development of active compliant lightweight mechanism structures. In: Proceedings of the 4th IFAC-symposium on mechatronic systems, Heidelberg, 12–14 Sept 2006. – Preprints on CD-ROM

Hufenbach W, Gude M, Modler N, Kirvel Ch (2007) Novel function-integrated lightweight solutions based on thermoplastic composites and material-adapted piezoceramic actor modules. Adv Mater Technol (AMT) 3–4:254–260

Mattson CA, Howell LL, Magleby SP (2004) Development of commercially viable compliant mechanisms using the pseudo-rigid-body model: case studies of parallel mechanisms. J Intell Mate Syst Struct 15:195. doi:10.1177/1045389X04033256

Midha A, Howell LL, Norton TW (2000) Limit positions of compliant mechanisms using the pseudo-rigid-body model concept. Mech Mach Theory 35:99–115

Modler N (2008) Nachgiebigkeitsmechanismen aus Textilverbunden mit integrierten aktorischen Elementen. Dissertation Technische Universität Dresden

Modler N, Hufenbach W, Modler K-H, Lovasz E-Ch, Perju D, Margineanu DA (2009) Design of compliant mechanisms with integrated actuators. In: SYROM 2009 – Proceedings of the 10th IFToMM international symposium on science of mechanisms and machines, Braşov (Romania), 12–15 October 2009, pp 655–664

Nashrul M, Zubir M, Shirinzadeh B, Tian Y (2009) A new design of piezoelectric driven compliant-based microgripper for micromanipulation. Mech Mach Theory 44:2248–2264, Accepted 8 July 2009, Available online 18 Aug 2009

Tanık E, Söylemez E (2010) Analysis and design of a compliant variable stroke mechanism. Mech Mach Theory 45:1385–1394

Electrical Performance of Electrically Conductive Silicone Rubber in Dependence on Tensile Load According to Various Parameters

M. Issa and L. Zentner

Abstract The paper presents the results of many investigations to make clear how the different parameters of tensile load which will possibly control the use of a strain sensor made of the conductive silicone rubber affect the electrical properties of this material. In order to clarify the behavior of this conductive silicone rubber during the cyclic load other tests are implemented on several specimens for each test to compare their behavior. The compliant systems with such sensors do not require mounting, and also they can be miniaturized. The target of these investigations is to use the conductive silicone rubber as a strain sensor by means of changing its electrical properties depending on the tensile load under different conditions.

Keywords Conductive silicone rubber • Compliance • Tensile test • Silicone elastomers • Cyclic load • Strain sensor

1 Introduction

The carbon-black filled silicone rubber -which is electrically conductive- has the ability to change its resistance as a result of deformation. These properties make this material suitable to develop force or deformation sensors (Valenta and Bojtos 2008). Another benefit of conductive silicone rubber is its high elasticity. This new silicone rubber sensor offers new opportunities for the measurement and control of high elasticity structures (Valenta and Huba 2007). Generally one can say that this material could be used as sensor for the measurement of mechanical forces and this sensor will be useful for example in the field of robotic applications (Patent application. Robust sensor for measuring mechanical force). A compliant gripper with sensor elements made of carbon-black filled silicone rubber, which were used to detect the presence of the object and its size is presented in

M. Issa (✉) • L. Zentner
Department of Mechanical Engineering, Mechanism Technology Group,
Technische Universität Ilmenau, Ilmenau, Germany
e-mail: mirna.issa@tu-ilmenau.de; lena.zentner@tu-ilmenau.de

E.-C. Lovasz et al. (eds.), *Microactuators and Micromechanisms*, Mechanisms and Machine Science 30, DOI 10.1007/978-3-319-15862-4_16

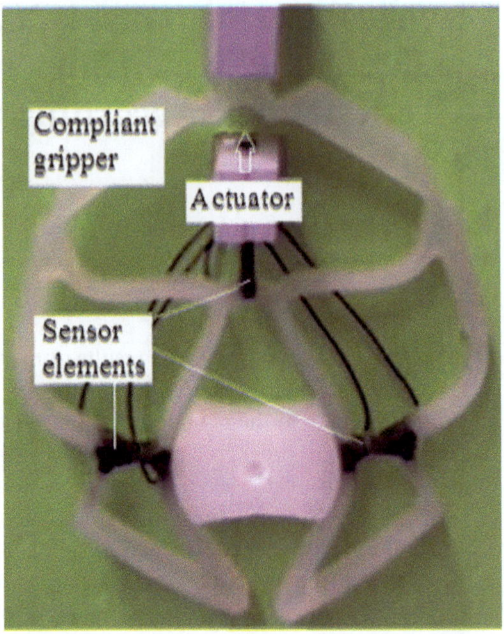

Fig. 1 Compliant gripper with embedded sensors made of conductive silicone rubber (Issa and Zentner 2012; Issa and Petkovic 2013)

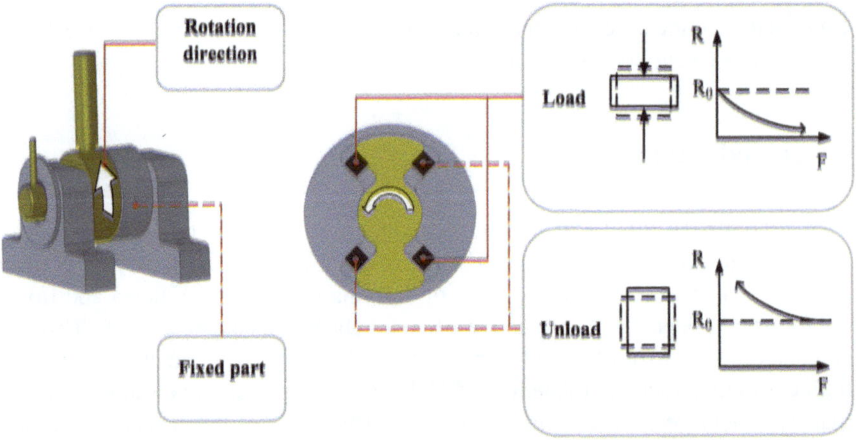

Fig. 2 Compliant joining structure with embedded sensors made of conductive silicone rubber (Issa and Petkovic 2011)

(Issa and Zentner 2012; Issa and Petkovic 2013). The sensor elements are parts of the compliant structure and also have the sensing ability (see Fig. 1).

Another application is a compliant joining structure with embedded sensors made of conductive silicone rubber (Issa and Petkovic 2011). The sensory characteristics are for detecting the rotation direction of the joint during the collision in order to prevent strong impacts or at least to minimize them, as shown in Fig. 2.

As it is our aim to design a strain sensor, different parameters for tensile tests are studied. The change of the electrical material properties is measured in dependence on different parameters – at first on the strain and then on the velocity of the tensile load – and also under cyclic load. The changing of the electrical resistance of the specimens is measured with the help of Cu-wire netting as electrodes. These electrodes are attached to both ends of the specimens during the vulcanization of the specimens to ensure good connections (Risto 2013).

2 Preparation of the Specimen

The specimen consists of five parts. The first one is made of electrically conductive silicone rubber (Elastosil® R 570/50 MH C1) with dimensions of $100 \times 2 \times 5\ mm^3$ and the other parts are made of non-conductive silicone rubber (Elastosil® R 420/70 MH C1) with each part of them being the dimensions of $10 \times 11.5 \times 5\ mm^3$. The five parts are connected by using a special technology (see Sect. 2.1). The material, which the specimens are made of, will be used for press moulding.

2.1 Connecting the Conductive Silicone Rubber with the Non-conductive Silicone Rubber

In order to test the electrically conductive silicone rubber under tensile load without any other effect like for instance the mechanical compressive forces, it is necessary to use a holder shown in Fig. 3 which needs an additional part made of silicon to stabilize the specimen in it. It is better to make this part of non-conductive material in order not to confuse the electrical resistance measurements. Therefore it is useful to connect the non-conductive silicone rubber with conductive silicone rubber. The

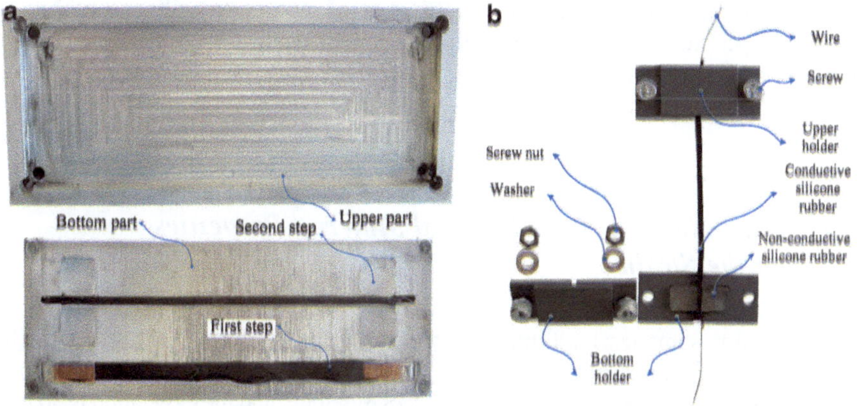

Fig. 3 (**a**) Press tool; (**b**) Holder (Issa and Zentner 2012)

holder is clamped in the jaws of a test machine. During the study, various connection modes of non-conductive silicone rubber with conductive silicone rubber are investigated. It was found that the best process for a good connection to press the prefabricated conductive silicone rubber with the raw non-conductive silicone rubber. In this way both materials are bonded together, and the connecting line is nearly a straight line, which cannot be achieved by another process. For example, there is no connection between the materials (conductive and non-conductive silicone rubber) if all parts (which are made of these materials) are prefabricated before they are pressed together. On the other hand, if all parts are made of raw material, the connecting line will not be straight, which can affect the electrical resistance of the specimens through different cross-sections.

2.2 Preparation of Tensile Specimens Using a Hot Press Machine

The press tool is designed to produce specimens with the help of the hot press machine. The design of this press tool enables it to produce the specimen in two steps, as shown in Fig. 3. We press the conductive silicone rubber specimen with the available Cu-wire netting attached to both ends of this specimen, and then the prefabricated conductive silicone rubber specimen is pressed again with the raw non-conductive silicone rubber. The Cu-wire netting is placed between them accordingly.

3 Measurements of the Electrical Material Properties as a Function of Tensile Load

The tests are performed by using a Zwick ProLine material-testing machine Z005. The tensile test is executed with different parameters to prove whether these parameters have an effect on the electrical resistance of the silicone rubber or not. All specimens are unloaded before testing, because each specimen needs time for relaxation after load before it can be used again, and this relaxation time depends on the maximum stretch (Spielmann 2006).

3.1 Dependence of the Electrical Material Properties on the Strain

In this section some tests are implemented to define the effect of the strain on the electrical material properties. The condition here is a constant velocity in all tests with its value being 10 mm/min. The specimen is 78 mm long, and the elongation is 30 mm

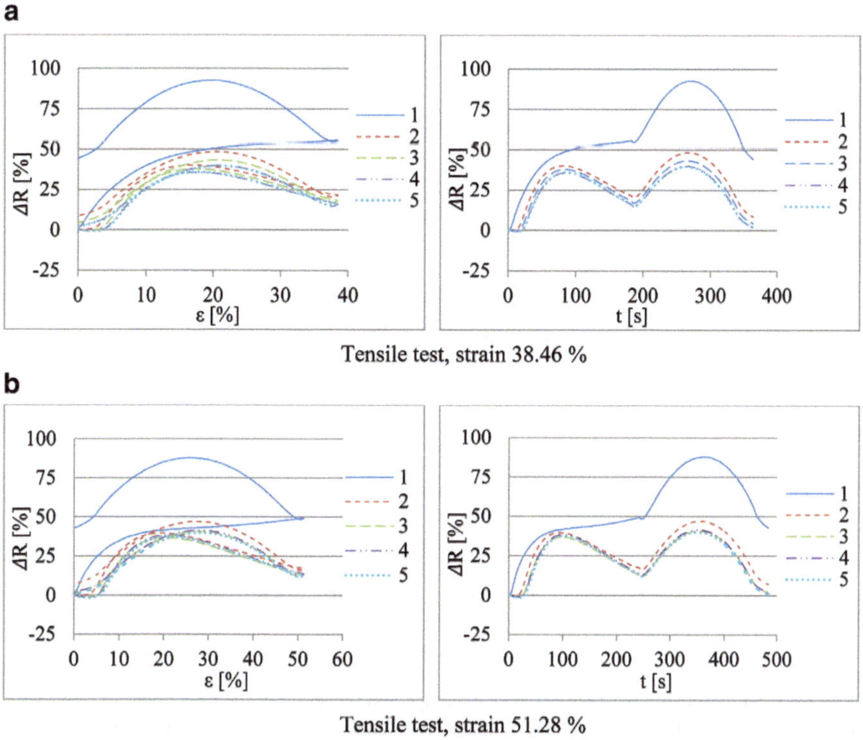

Fig. 4 The relationship between resistance change and the strain or the time under tensile test at different strains and for several times: (**a**) Strain 38.46 %; (**b**) Strain 51.28 %

(38.46 %) in the first test; in the second one it is 40 mm (51.28 %). Each test is repeated five times immediately after each other. The tensile test is illustrated in Fig. 4.

In the tests, it becomes clear that the specimen must be pretested because the curves at the first load and at the following four loads differ qualitatively. The range in which the resistance at different strain values changes is almost the same for both investigations with 30 mm and 40 mm elongation. The change of resistance in the first test reaches a value of about 90 % in both cases, and it reaches almost half the value (45–50 %) in the last 4 tests in these two investigations. Furthermore, one cannot detect the strain value via the resistance value, because there are two values of the strain for each value of the resistance, as clarified in the conclusion later.

3.2 Dependence of Electrical Material Properties on the Velocity of the Tensile

The purpose of these tests is to determine the effect of the velocity of the tensile load on the electrical material properties. The applied elongation during the tensile

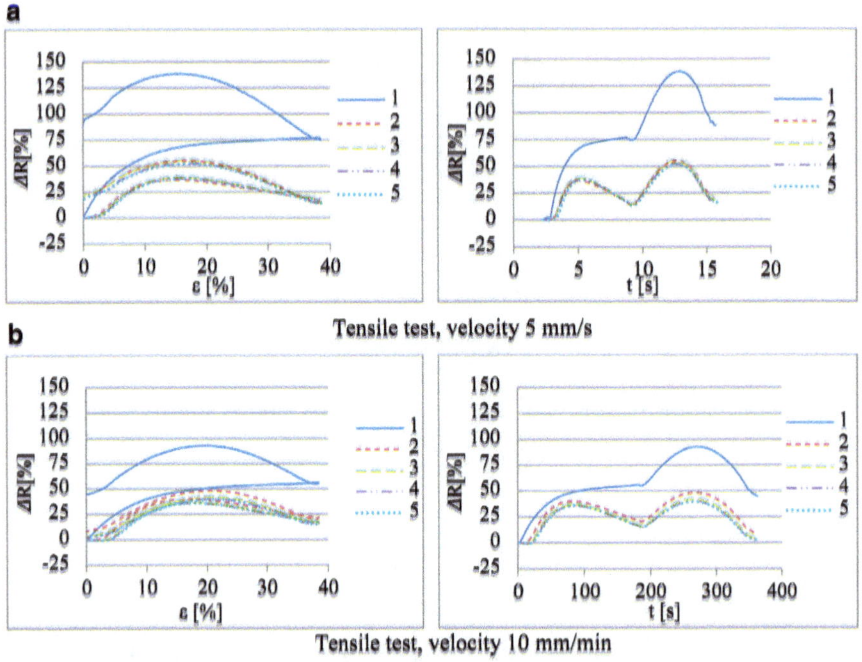

Fig. 5 The relationship between resistance change and the strain or the time under tensile load at different velocity values and for several times: (**a**) Velocity 5 mm/s; (**b**) Velocity 10 mm/min

load is 30 mm (38.46 %) and this value is still constant in all tests, but the velocity is different: in the first test it is 5 mm/s and in the second test it is 10 mm/min, as shown in Fig. 5.

Figure 5 illustrates that the faster the load on the specimen is, the greater is the maximum value of the resistance change, and that is clear in the first cycle, in the following four cycles, the difference between both tests is not so great.

3.3 Change of the Electrical Material Properties Under Cyclic Load

The cyclic load test is executed with the purpose of spotting the behavior of the conductive silicone rubber under these specific test conditions and to illustrate the change of its electrical properties caused by the deformation. The test consists of 16 cycles, at every cycle the specimen is stretched 30 mm (38.46 %) and then unloaded until the starting point of this test. The velocity of the test is 10 mm/min. At the end of this test it is found that the specimen is sagged, as shown in Fig. 6.

In order to cope with the aforementioned problem another cyclic test is executed. This test consists of 16 cycles. During every cycle the specimen is stretched 30 mm

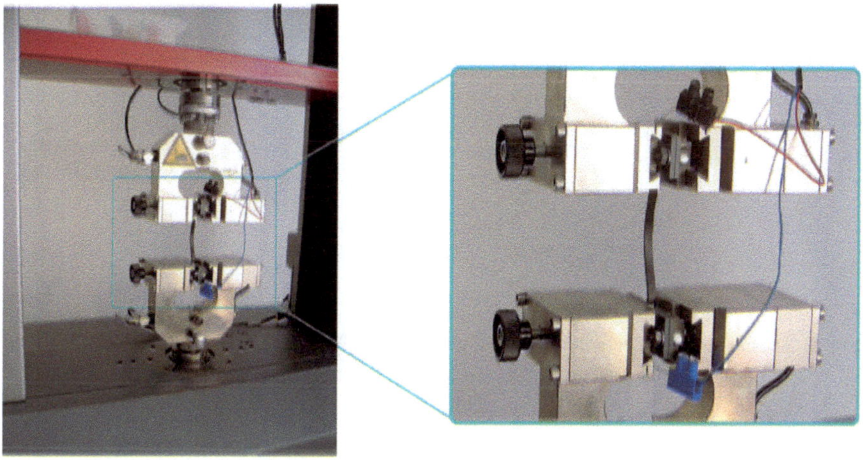

Fig. 6 Sagged specimen after cyclic load

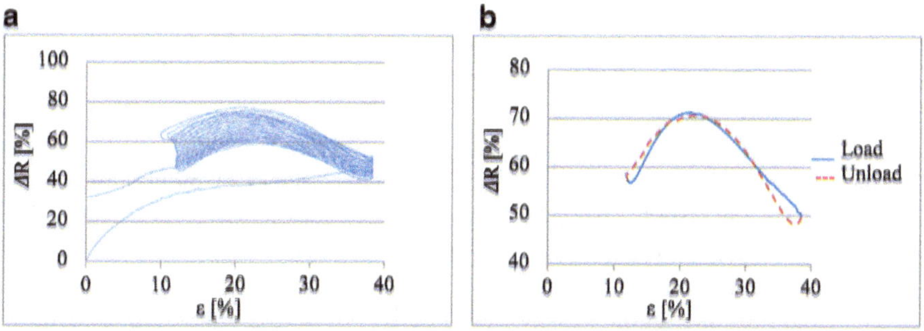

Fig. 7 Tensile test: (**a**) Cyclic load, unload until force equal 0 N. (**b**) 6th cycle

(38.46 %) and then unloaded until a tensile force equal to zero and this end point of the unloading step is set by the machine. This test has the same velocity of 10 mm/min. The results are illustrated in Fig. 7.

At the beginning of this test the sagging increases after every cycle, and it remains almost constant up the fourth cycle. That means, at each unloading step of these four cycles the force is zero at different strain percentages and the curves do not have the same starting point. After many executed tests it was found that the force value is zero in most cases at the point of about 7 mm and this is equal to 8.97 % of the length of the specimen. In order to make the starting point the same with all curves for easy comparison between them, it is important to unload the specimen at each unloading step until a certain strain percentage, which is set at 7 mm in the following test, as illustrated in Fig. 8. The sixth cycle of the cyclic load shows the run of the curve during the loading and the unloading step clearly, as clarified in Figs. 7b and 8b.

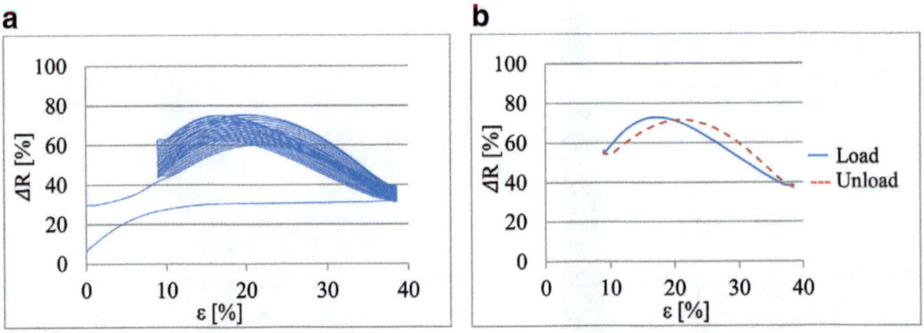

Fig. 8 Tensile test: (**a**) Cyclic load, unload until strain 8.97 %; (**b**) The 6th cycle

4 Conclusions

In this section, the results of the above-mentioned tests as well as the rules for application of conductive silicone rubber to develop a sensor element are presented as following:

1. The first tensile test applied on each element for any parameters is different from other tests as seen in Figs. 4 and 5. Hence the actuation of the elements for several times is necessary for application as a sensor.
2. The resistance change of the conductive silicone rubber element significantly depends on the speed especially during the first tensile load. As illustrated in Fig. 5, conductive silicone rubber is suitable for applications as a sensor with different measurement speeds (the tested one at least).
3. The length of the sagging -where the force is zero during unloading or loading- depends on for how many times the element was loaded and also on the amount for which it was stretched. In the cyclic loading (at 38.46 % strain each time), it is noticed that after the fourth cycle the sagging is almost constant (see Fig. 7). Consequently the sensor has a better behavior under cyclic load.
4. The element must be stretched before using, so that we can overcome the sagging which occurs under cyclic load. If the force acts no longer, the element will be relaxed to the selected strain. The run of the curve of resistance change is nearly the same for many cycles, as shown in Fig. 8. Thus the qualitative run of the curve can be used for each sensor element.
5. In the diagram showing the change of resistance depending on strain (e.g. at 38.46 %) one cannot determine how much the elongation will be at a point whose resistance value is known, even though it is clear when the loading or the unloading step in this diagram occurs, because there are two values of strain for each value of resistance, as shown in Fig. 9. If one takes a smaller strain (e.g. 6.41 %, it is 5 mm), one can determine the elongation. Therefore, application for small strain loading is recommended.

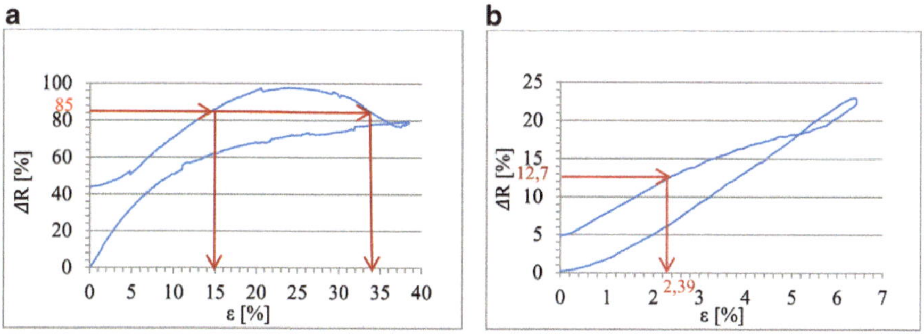

Fig. 9 The difference of behavior between two different applied strains. (**a**) Tensile test with strain 38.46 %. (**b**) Tensile test with strain 6.41 %

5 Summary

The results of the above-mentioned investigations show that this material can be used as sensor element in the application for small strain loading, and where only a qualitative analysis important is. Furthermore, this material is suitable to use as sensor element in the compliant mechanisms, where the compliance has to be insured. In addition to that, they do not require mounting, and also they can be miniaturized.

References

Issa M, Petkovic D (2011) Embedded-sensing elements made of conductive silicone rubber for compliant robotic joint. In: 56th Internationales Wissenschaftliches Kolloquium TU Ilmenau

Issa M, Petkovic D (2013) Sensor elements made of conductive silicone rubber for passively compliant gripper. Adv Manuf Technol 69:1527–1536

Issa M, Zentner L (2012) Sensorelemente aus leitfähigem silikon für einen nachgiebigen greifer. In: Mechanismentechnik 2012, Workshop Ilmenau-Budapest-Niš

Patent application. Robust sensor for measuring mechanical force. Inventor: Karsten Weiß, Pat. Nr. DE 197 50 671A1

Risto U (2013) Zur Charakterisierung und Anwendung des Durchschlagverhaltens von nachgiebigen rotationssymmetrischen Strukturen. PhD thesis, TU Ilmenau

Spielmann C (2006) Entwicklung eines Sensors für einen ferrofluidischen Flächenaktor. Diploma thesis

Valenta L, Bojtos A (2008) Mechanical and electrical testing of electrically conductive silicone rubber. Trans Tech Publications, Aedermannsdorf, pp 179–184

Valenta L, Huba A (2007) High elasticity strain gauge made from conductive silicone rubber, materials science, testing and informatics. Trans Tech Publications, Aedermannsdorf, pp 709–716

MIX
Papier aus verantwortungsvollen Quellen
Paper from responsible sources
FSC® C105338

If you have any concerns about our products,
you can contact us on
ProductSafety@springernature.com
In case Publisher is established outside the EU,
the EU authorized representative is:
Springer Nature Customer Service Center GmbH
Europaplatz 3, 69115 Heidelberg, Germany
Printed by Libri Plureos GmbH
in Hamburg, Germany